Basic Physical Chemistry Calculations

Basic Physical Chemistry Calculations

H. E. Avery, B.Sc., Ph.D. and D. J. Shaw, B.Sc., Ph.D.

*Department of Chemistry
and Metallurgy,
Lanchester Polytechnic*

*Department of Chemistry
Liverpool Polytechnic*

LONDON
BUTTERWORTHS

THE BUTTERWORTH GROUP

ENGLAND

Butterworth & Co (Publishers) Ltd
London: 88 Kingsway, WC2B 6AB

AUSTRALIA

Butterworth Pty Ltd
Sydney: 586 Pacific Highway, NSW 2067
Melbourne: 343 Little Collins Street, 3000
Brisbane: 240 Queen Street, 4000

CANADA

Butterworth & Co (Canada) Ltd
Toronto: 14 Curity Avenue, 374

NEW ZEALAND

Butterworths of New Zealand Ltd
Wellington: 26 28 Waring Taylor Street, 1

SOUTH AFRICA

Butterworth & Co (South Africa) (Pty) Ltd
Durban: 152 154 Gale Street

B/541.301

First published 1971
Reprinted with corrections 1973
© Butterworth & Co (Publishers) Ltd 1971
ISBN 0 408 70046 7 Standard
      0 408 70047 5 Limp

Printed in England by Chapel River Press, Andover, Hants

Contents

Preface

The purpose of this book is to illustrate methods of approach towards solving numerical problems encountered in Physical Chemistry and to facilitate a general understanding of this subject. The book should prove to be suitable for students studying at the levels of 1st and 2nd year B.Sc., Higher National Certificate, Higher National Diploma and Part 1 G.R.I.C. A companion volume entitled *Advanced Physical Chemistry Calculations* covers material appropriate to the levels of Honours B.Sc. and Part 2 G.R.I.C.

The first chapter consists of a discussion of Système International (SI) units and of physico-chemical methods of expression. The next six chapters are devoted to various branches of Physical Chemistry and consist of a series of worked examples followed by additional examples to be solved by the reader. In the worked examples, theoretical discussion is kept to a reasonable minimum. Answers to the additional examples are provided and, in the few cases where there is no corresponding worked example, a brief directive towards the method of solution is given with the answer. Values of commonly encountered physical constants and conversion factors are given in Appendices I and II rather than with each individual question. The final chapter consists of a number of multiple choice and true-false questions, designed to test the reader's awareness of some of the basic concepts in Physical Chemistry. Some of these questions have been prompted by examiners' criticisms of past performances in Royal Institute of Chemistry examinations.

Some of the calculations are taken from past examination papers, and in this respect we wish to thank the Royal Institute of Chemistry, the Universities of Birmingham, Bristol, Durham, Leeds, Liverpool, Manchester, Nottingham, Salford and Sheffield, and Liverpool Polytechnic for permission to publish. Where necessary, we have amended the original questions so as to conform with SI units and the physico-chemical notation outlined in Chapter 1.

Finally, we wish to thank Dr. A. R. Denaro, Dr. L. F. Moore

and Dr. A. L. Smith for a number of helpful suggestions, and also our wives for their help in preparing the manuscript and checking the text.

Liverpool   H. E. AVERY D. J. SHAW

CHAPTER 1

Units and Methods of Expression in Physical Chemistry

A good case can be made in favour of a single international language to replace the many national and regional languages in present use; however, for various reasons, the prospect of such a change taking place in the forseeable future is extremely remote. A much stronger case can be made in favour of an international scientific and technological language in which the adopted methods of expression avoid any possible ambiguity as to their meaning and in which physical quantities are expressed, as far as possible, in relation to a single coherent set of units. Such a language (albeit not, as yet, entirely perfected) now exists in physical science, and is outlined in this chapter in so far as it is relevant to the contents of this book. (For a more detailed account than that given in this chapter, see M. L. McGlashan, *Physico-Chemical Quantities and Units*, Royal Institute of Chemistry, Monographs for Teachers No 15, 1968.)

SI Units

Since our system of pure numbers is decimal, the manipulation and recording of physical quantities is facilitated if the alternative units in which a particular physical quantity might be expressed differ by factors of ten. Moreover, matters are simplified further

1

if a single system of such units is adopted. An international system of units has now been formulated. It is referred to as SI (abbreviation for Système International d'Unités).

SI is based on the following seven independent physical quantities (see *Table 1.1*).

Table 1.1

Physical quantity	Symbol(s)	Basic SI unit	Unit symbol
length	l, b, d, h, r, s, etc.	metre	m
mass	m	kilogramme	kg
time	t	second	s
electric current	I	ampere	A
thermodynamic temperature	T	kelvin	K
amount of substance*	n	mole	mol
luminous intensity	I_v	candela	cd

*See section on molar quantities, page 15.

[*Note*. A reason for selecting metre, kilogramme and second (m.k.s.) as the basic units of mass, length and time, rather than, for example, centimetre, gramme and second (c.g.s.) is so that the unit of work (mass×acceleration×distance) is the joule ($kg\ m^2\ s^{-2}$), the magnitude of which is more convenient than that of the erg ($g\ cm^2\ s^{-2} = 10^{-7}$ joule). The unit of electric current, the ampere, is so defined that the unit of electric energy (potential difference ×current×time) is also the joule.]

(Electric charge or electric potential difference could, alternatively, have been chosen as a basic quantity instead of electric current; this is apparent from the definitions of their units.)

SI also includes two fundamental quantities with *dimensionless units* (see *Table 1.2*).

Table 1.2

Physical quantity	Symbols	SI unit	Unit symbol
plane angle	$\alpha, \beta, \gamma, \theta, \phi$	radian	rad
solid angle	ω, Ω	steradian	sr

Other physical quantities can be expressed in SI units which are derived from the above basic units by appropriate multiplication, division, integration and/or differentiation without the introduction of any numerical factors (including powers of ten). In this sense, SI is described as a coherent system of units. A selection of derived SI units, some of which have been given special names, is given in *Tables 1.3* and *1.4*.

Table 1.3

DERIVED SI UNITS WITH SPECIAL NAMES

Physical quantity	Symbol(s)	Name of SI unit	Unit symbol	Definition
frequency	v, f	hertz	Hz	s^{-1}
force	F	newton	N	$kg\ m\ s^{-2} = J\ m^{-1}$
energy (all forms)*	E, U, V, etc.	joule	J	$N\ m = kg\ m^2\ s^{-2} = C\ V = V\ A\ s$
power	P	watt	W	$J\ s^{-1} = kg\ m^2 s^{-3} = V\ A$
electric charge	Q	coulomb	C	$A\ s$
electric potential difference	$E, \psi, \zeta,$ ϕ, η, etc.	volt	V	$J\ A^{-1} s^{-1} = kg\ m^2\ s^{-3}\ A^{-1}$
electric resistance	R	ohm	Ω	$V\ A^{-1} = kg\ m^2\ s^{-3}\ A^{-2}$
electric capacitance	C	farad	F	$C\ V^{-1} = A^2\ s^4\ kg^{-1}\ m^{-2}$
inductance	L, M	henry	H	$V\ A^{-1} s = kg\ m^2\ s^{-2}\ A^{-2}$
magnetic flux	Φ	weber	Wb	$V\ s = kg\ m^2\ s^{-2}\ A^{-1}$
magnetic induction	B	tesla	T	$Wb\ m^{-2} = kg\ s^{-2}\ A^{-1}$
luminous flux	Φ	lumen	lm	$cd\ sr$
illumination	E	lux	lx	$lm\ m^{-2} = cd\ sr\ m^{-2}$

* See *Note* on page 2.

Note. (1) Quantity symbols are always printed in italic (sloping) type (bold-faced italic for vector quantities). Unit symbols are printed in roman (upright) type.

(2) Full stops are not used between units to represent multiplication, but units are spaced to avoid possible confusion, such as between $1\ m\ s^{-1}$ (1 metre per second) and $1\ ms^{-1}$ (which could be misinterpreted as 1 reciprocal millisecond). $10^3\ s^{-1}$ is an unambiguous representation of the latter quantity.

(3) Plural forms to unit symbols are not used, e.g. 10 kg and not 10 kgs.

Table 1.4

SOME OTHER SI UNITS AND RECOMMENDED QUANTITY SYMBOLS

(A number of points which are particularly relevant to the quantities marked with an asterisk are discussed later in this chapter.)

Physical quantity	Symbol(s)	SI unit (if any)
area: $A = \int l\,db$	A	m^2
volume: $V = \int A\,dh$	V	m^3
velocity: $u = ds/dt$	u, v, c	$m\ s^{-1}$
angular velocity: $\omega = d\theta/dt$	ω	$rad\ s^{-1}$
momentum: $P = mu$	p	$kg\ m\ s^{-1} = N\ s$
angular momentum: $L = rp$	L	$kg\ m^2\ s^{-1} = J\ s$
acceleration: $a = du/dt$	a, g (free fall)	$m\ s^{-2}$
moment of inertia: $I_z = \int (x^2+y^2)\,dm$	I	$kg\ m^2$
weight	$G, (W)$	N
density: $\varrho = m/V$	ϱ	$kg\ m^{-3}$
*specific volume: $v = V/m$	v	$m^3\ kg^{-1}$
pressure†	p	$N\ m^{-2} = J\ m^{-3}$ $= kg\ m^{-1}\ s^{-2}$
shear stress	τ	$N\ m^{-2}$
dynamic viscosity: $\eta = \tau_{zz}/(du_z/dz)$	η	$kg\ m^{-1}\ s^{-1}$
kinematic viscosity: $v = \eta/\varrho$	v	$m^2\ s^{-1}$
diffusion coefficient	D	$m^2\ s^{-1}$
surface tension	γ, σ	$N\ m^{-1} = J\ m^{-2}$ $= kg\ s^{-2}$
molecular mass	m	kg
*molar mass: $M = m/n$	M	$kg\ mol^{-1}$
*molar volume: $V_m = V/n$	V_m	$m^3\ mol^{-1}$
*relative atomic mass	A_r	dimensionless
*relative molecular mass	M_r	dimensionless
work: $w = Fs$	w	J
quantity of heat	q	J
thermodynamic (internal) energy	U	J
enthalpy: $H = U+pV$	H	J
Helmholtz free energy function: $A = U-TS$	A	J
Gibbs free energy function: $G = H-TS = U+pV-TS$	G	J

† Although not, as yet, officially adopted as a specially named SI unit, the name pascal (symbol, Pa) is sometimes used to denote the SI unit of pressure, $N\ m^{-2}$.

Table 1.4 *(cont.)*

Physical quantity	Symbol (s)	SI unit (if any)
entropy	S	J K^{-1}
heat capacity: $C_V = (\partial U/\partial T)_V$		
$C_p = (\partial H/\partial T)_p$	C_V, C_p	J K^{-1}
molar gas constant: $R = N_A k$	R	J K^{-1} mol^{-1}
Boltzmann constant: $k = R/N_A$	k	J K^{-1}
*Avogadro constant	N_A, L	mol^{-1}
*concentration of substance B:		
$c_B = n_B/V$	c_B, [B]	mol m^{-3}
*molality of solute B:		
$m_B = n_B/n_A M_A$	m_B	mol kg^{-1}
mole fraction of substance B:		
$x_B = n_B/\sum n_B$	x_B	dimensionless
stoichiometric coefficient of substance B		
(+ ve for products, $-$ ve for reactants)	ν_B	dimensionless
degree of dissociation	α	dimensionless
*molar thermodynamic energy:		
$U_m = U/n$	U_m	J mol^{-1}
*molar enthalpy: $H_m = H/n$	H_m	J mol^{-1}
*molar free energy: $A_m = A/n$,		
$G_m = G/n$	A_m, G_m	J mol^{-1}
*molar entropy: $S_m = S/n$	S_m	J K^{-1} mol^{-1}
*molar heat capacity: $C_{V, m} = C_V/n$,		
$C_{p, m} = C_p/n$	$C_{V, m}, C_{p, m}$	J K^{-1} mol^{-1}
partial molar volume of substance B:		
$V_B = (\partial V/\partial n_B)_{T, p, n_c \dots}$	V_B	m^3 mol^{-1}
chemical potential of substance B		
$\mu_B = (\partial G/\partial n_B)_{T, p, n_c \dots}$	μ_B	J mol^{-1}
fugacity	p^*	N m^{-2}
*absolute activity of substance B:		
$\lambda_B = \exp(\mu_B/RT)$	λ_B	dimensionless
*relative activity of solvent A:		
$a_A = \lambda_A/\lambda_A$●†	a_A	dimensionless
*relative activity of solute B:		
$a_B = m_B\gamma_B/m$⊖†	a_B	dimensionless
*activity coefficient of solute B:		
$\gamma_B = (\lambda_B/m_B)/(\lambda_B/m_B)^\infty$†	γ_B	dimensionless
$y_B = (\lambda_B/c_B)/(\lambda_B/c_B)^\infty$†	y_B	dimensionless
osmotic pressure of a solution	Π	N m^{-2}
osmotic coefficient of solvent A	g, ϕ	dimensionless

† ● indicates a pure substance; ⊖ indicates a standard value of a property (e.g. 1 atm, 1 mol kg^{-1}); ∞ indicates a limiting value at infinite dilution.

Table 1.4 (*cont.*)

Physical quantity	Symbol(s)	SI unit (if any)
*equilibrium constant: $K_p = \Pi(p_B)^{\nu_B}$	K_p	$(\text{N m}^{-2})^{\Sigma_i \nu_B}$
*equilibrium constant: $K_c = \Pi(c_B)^{\nu_B}$	K_c	$(\text{mol m}^{-3})^{\Sigma \nu_B}$
*equilibrium constant: $K_m = \Pi(m_B)^{\nu_B}$	K_m	$(\text{mol kg}^{-1})^{\Sigma_i \nu_B}$
*equilibrium constants		
$K_p*/_{p\ominus} = \Pi\left(\dfrac{p_B^*}{p^\ominus}\right)^{\nu_B}$,	$K_p*/_{p\ominus}$,	
$K_{m\gamma/m\ominus} = \Pi\left(\dfrac{m_B\gamma_B}{m^\ominus}\right)^{\nu_B}$, etc.	$K_{m\gamma/m\ominus}$, etc.	dimensionless
partition functions	Q, q	dimensionless
electric field strength	E	V m^{-1}
magnetic field strength	H	A m^{-1}
charge density: $\varrho = Q/V$	ϱ	C m^{-3}
surface charge density: $\sigma = Q/A$	σ	C m^{-2}
electric current density	j	A m^{-2}
*permittivity	ε	$\text{F m}^{-1} = \text{kg}^{-1}$ $\text{m}^{-3}\,\text{s}^4\,\text{A}^2$
*relative permittivity‡: $\varepsilon_r = \varepsilon/\varepsilon_0$ (ε_0 = permittivity of vacuum)	ε_r	dimensionless
*permeability	μ	$\text{H m}^{-1} = \text{kg m}$ $\text{s}^{-2}\,\text{A}^{-2}$
*relative permeability: $\mu_r = \mu/\mu_0$ (μ_0 = permeability of vacuum)	μ_r	dimensionless
polarizability of a molecule	α	$\text{C m}^2\,\text{V}^{-1}$
dipole moment of a molecule	μ, p	C m
elementary charge (proton or electron)	e	C
charge number of ion i (+ve or −ve)	z_i	dimensionless
Faraday constant: $F = N_A e$	F	C mol^{-1}
ionic strength: $I = \frac{1}{2}\sum m_i z_i^2$	I	mol kg^{-1}
*mean ionic activity coefficient of an electrolyte: $\gamma_\pm = (\gamma_+^{\nu+}\gamma_-^{\nu-})^{1/(\nu_+ + \nu_-)}$	$\gamma_\pm$	dimensionless
*electric conductance: $G = 1/R$	G	Ω^{-1}
*electric conductivity: $\varkappa = j/E$	$\varkappa$	$\Omega^{-1}\,\text{m}^{-1}$
*molar conductivity: $\Lambda = \varkappa/c$	Λ	$\Omega^{-1}\,\text{m}^2\,\text{mol}^{-1}$
transport number of ion i	t_i	dimensionless
mobility of a charged particle $u = v/E$	u	$\text{m}^2\,\text{s}^{-1}\,\text{V}^{-1}$
rate of increase of concentration of substance B	$v_B, \dfrac{dc_B}{dt}, \dfrac{d[B]}{dt}$	$\text{mol m}^{-3}\,\text{s}^{-1}$
rate constant of a $(n+1)$th order reaction	k, k_r	$\text{m}^{3n}\,\text{mol}^{-n}\,\text{s}^{-1}$

‡ Also called dielectric constant (symbol, D) when it is independent of E.

Table 1.4 *(cont.)*

Physical quantity	Symbol (s)	SI unit (if any)
activation energy of a reaction	ΔE, $\Delta E^{\ddagger}$	J mol^{-1}
collision number	Z	m^{-3} s^{-1}
quantum yield	Φ	dimensionless
quantum numbers	J, v, etc.	dimensionless
refractive index	n	dimensionless
Planck's constant	h	J s
wavelength	λ	m
*wave number: $\bar{v} = 1/\lambda$	$\bar{v}$, σ	m^{-1}
transmittance: $T = I/I_0$	T	dimensionless
decadic absorbance (decadic extinction)§:		
$A = -\log_{10} T$	A	dimensionless
molar Napierian extinction coefficient	$\varkappa$	m^2 mol^{-1}
molar decadic extinction coefficient:		
$\varepsilon = A/lc$	ε	m^2 mol^{-1}

§ Formerly called optical density.

(4) The use of a single solidus (/) for expressing derived units (e.g. m s^{-1}/V m^{-1} for mobility rather than m^2 s^{-1} V^{-1}) is permissible, but better avoided. More than one solidus in the same expression (e.g. m/s/V/m for mobility) should never be used owing to the ambiguity which is created.

PREFIXES FOR SI UNITS

SI permits the use of the following prefixes *(Tables 1.5 and 1.6)* to denote decimal fractions and multiples of basic SI units and derived SI units with special names. To avoid a multiplicity of prefixes, the factors are 10^{3n} except around unity where additional prefixes are available to denote 10^{-2}, 10^{-1}, 10 and 10^2. Compound prefixes should not be used.

These prefixes are, of course, not absolutely necessary, but are convenient in so far as they avoid the use of inconveniently large and small numerical values and sometimes facilitate appreciation of the magnitudes of physical quantities, as illustrated by the following examples.

Table 1.5		
Factor	*Prefix*	*Symbol*
10^{-1}	deci	d
10^{-2}	centi	c
10^{-3}	milli	m
10^{-6}	micro	μ
10^{-9}	nano	n
10^{-12}	pico	p
10^{-15}	femto	f
10^{-18}	atto	a

Table 1.6		
Factor	*Prefix*	*Symbol*
10	deka	da
10^2	hecto	h
10^3	kilo	k
10^6	mega	M
10^9	giga	G
10^{12}	tera	T

(1) The collision diameter of a nitrogen molecule is $3{\cdot}75\times10^{-10}$ m or 0·375 nm (nanometre).

(2) The concentration of an aqueous solution of potassium chloride is 100 mol m^{-3} or 0·1 mol dm^{-3}. 0·1 mol dm^{-3} is the more convenient of these alternatives because, (*a*) 1 m^3 of solution is in considerable excess of the scale of a typical laboratory experiment, (*b*) it is numerically the same as the now obsolete term, molar concentration (mol l^{-1}), and (*c*) it is numerically similar to the molality expressed in mol kg^{-1}.

Since SI is a coherent system of units, it follows that, when performing numerical calculations, no thought about conversion factors is necessary if physical quantities are expressed in unprefixed SI units, as illustrated by the following examples.

(1) Use the relationship, $E = N_A hc\bar{v}$, to calculate the molar energy which corresponds to a wave number, $\bar{v}$, of 1 cm^{-1}. Substituting $N_A = 6{\cdot}022\,5\times10^{23}$ mol^{-1}, $h = 6{\cdot}625\,6\times10^{-34}$ J s, $c = 2{\cdot}997\,9\times10^8$ m s^{-1} and $\bar{v} = 10^2$ m^{-1}, automatically gives the value of E in J mol^{-1}, i.e.

$$E = 6{\cdot}022\,5\times10^{23}\times6{\cdot}625\,6\times10^{-34}\times2{\cdot}997\,9\times10^8\times10^2$$
$$(\text{mol}^{-1})\,(\text{J s})\,(\text{m s}^{-1})(\text{m}^{-1}) = 11{\cdot}962 \text{ J mol}^{-1}$$

(2) Calculate the potential, ψ, such that the relationship, $ze\psi = kT$, is valid at 25°C. Substituting $e = 1{\cdot}60\times10^{-19}$ C, $k = 1{\cdot}38\times10^{-23}$ J K^{-1}, $T = 298$ K and the appropriate value for z (dimensionless), automatically gives the value of ψ in volts, i.e.

$$\psi = \frac{1{\cdot}38\times10^{-23}\times298}{z\times1{\cdot}60\times10^{-19}}\,\frac{\text{J K}^{-1}\text{ K}}{\text{C}} = \frac{0{\cdot}025\,7}{z}\text{ V}$$

When using prefixed SI units in a numerical calculation, a conversion factor of an appropriate power of ten may be necessary to obtain the final answer in the desired unit. If the data for a numerical calculation is presented in prefixed SI units or non-SI units, the risk of miscalculation will often be minimized if conversion to non-prefixed SI units is made before substitution into the appropriate equation(s).

Non—SI Units

The following two tables *(1.7* and *1.8)* list a selection of units which are exactly defined in terms of SI but which are not coherent with SI. The units in the first table are decimal fractions or multiples of the corresponding SI unit and those in the second table are non-decimal with respect to the corresponding SI unit.

With the exception of the atmosphere (which must be retained in Physical Chemistry in view of its role as a standard state), these non-SI units are unnecessary and most of them must be regarded as obsolete (see first footnote to *Table 1.7*).

Some non-SI units, such as the minute, hour, degree Celsius and, possibly, the litre, will, of course be retained for everyday usage. The degree Celsius will doubtless continue to be used in some branches of physical science as a colloquialism for the thermodynamic temperature in excess of 273·15 K.

Table 1.7

Physical quantity	Name of unit	Unit symbol	Definition
length	ångström*	Å	10^{-10} m $= 10^{-1}$ nm
length	micron	$\mu^{\dagger}$	10^{-6} m $= \mu$m
length	millimicron	m$\mu^{\dagger}$	10^{-9} m $=$ nm
volume	litre	l	10^{-3} m^3 $=$ dm^3
force	dyne	dyn	10^{-5} N
energy	erg	erg	10^{-7} J
pressure	bar	bar	10^5 N m^{-2}
dynamic viscosity	poise	P	10^{-1} kg m^{-1} s^{-1}
concentration	'molar'‡	M	mol dm^{-3}
magnetic flux	maxwell	Mx	10^{-8} Wb
magnetic flux density (magnetic induction)	gauss	G	10^{-4} T

* In view of its convenience with respect to molecular dimensions, a number of scientists urge retention of the ångström as a recognized unit of length.

† μ means 10^{-6} and not 10^{-6} m; mμ, at best, means n or 10^{-9} and not 10^{-9}m.

‡ See page 16.

Table 1.8

Physical quantity	Name of unit	Unit symbol	Definition
energy	thermochemical calorie§	cal	4.184 J
pressure	atmosphere	atm	$1.013\ 25 \times 10^5$ N m^{-2}
pressure	conventional millimetre of mercury	mmHg	$13.595\ 1 \times 9.806\ 65$ N m^{-2} $= 133.322\ 39$ N m^{-2}
pressure	torr	Torr	$\dfrac{1.013\ 25 \times 10^5}{760}$ N m^{-2} $= 133.322\ 37$ N m^{-2}
temperature	degree Celsius**	°C	$T/°C = T/K - 273.15$††

§ Unlike the joule, the calorie is an ambiguous unit without further specification, e.g. 1 international calorie $= 1.000\ 67$ thermochemical calories.

** Celsius not Centigrade.

†† See page 12 for explanation of this type of notation.

It may be convenient to express experimental data in non-SI units in view of the method of measurement. Non-SI units should, however, in general, be avoided when recording physical quantities which have been calculated from experimental data. For example, if a mercury manometer is used for the purpose of investigating the pressure variation which accompanies a second-order gas phase reaction at constant temperature and volume taking place over a period of several minutes, it is convenient, and quite permissible, to tabulate pressures in mmHg or torr with the corresponding reaction times in min; however, the calculated rate constant should be expressed in units such as N^{-1} m^2 s^{-1} or mol^{-1} dm^3 s^{-1}, but not in torr^{-1} min^{-1}, etc.

ELECTRICAL AND MAGNETIC QUANTITIES AND UNITS— RATIONALIZATION

In SI electrical and magnetic units are based on the fundamental units of metre, kilogramme, second and ampere, and equations are rationalized.

The force, F, between charges, Q_1 and Q_2, separated by a distance, r, in a medium of permittivity, ε, is given by the rationalized expression,

$$F = \frac{Q_1 Q_2}{4\pi\varepsilon r^2}$$

Since Q has the dimension [current] [time], ε has the dimension [length]$^{-3}$ [mass]$^{-1}$ [time]^4 [current]^2. The permittivity of a vacuum, ε_0, in accordance with the above equation is equal to $8\cdot854\ 2\times10^{-12}$ kg^{-1} m^{-3} s^4 A^2.

The *relative permittivity* of a medium, $\varepsilon/\varepsilon_0$, can be referred to as the *dielectric constant*, D, of the medium if it is independent of electric field strength.

The above equation differs from the traditional Coulomb inverse square law equation by the inclusion of the factor, 4π, and is said to be rationalized. The reason for including 4π is so that its occurrence or otherwise in derived expressions might be in accordance with geometric expectation and not vice versa. For example, the equation for a parallel plate condenser, where the occurrence of 4π would not be expected from geometric considerations, is $\sigma = \dfrac{\varepsilon U}{4\pi d}$ in non-rationalized form and $\sigma = \dfrac{\varepsilon U}{d}$ in rationalized form; whereas the equation for an isolated spherical condenser, where the occurrence of 4π is expected from geometric considerations, is $Q = \varepsilon a U$ in non-rationalized form and $Q = 4\pi\varepsilon a U$ in rationalized form.

The force, F, between electric currents, I_1 and I_2, in parallel conductors of length, l, separated by a distance, d, in a medium of permeability, μ, is given by the rationalized expression,

$$F = \frac{2\mu I_1 I_2 l}{4\pi d}$$

Permeability, therefore, has the dimension, [length] [mass] [time]$^{-2}$ [current]$^{-2}$, and the permeability of a vacuum, μ_0, in accordance with the above equation is equal to $4\pi\times10^{-7}$ kg m s^{-2} A^{-2}.

ε_0 and μ_0 are related by the expression, $\varepsilon_0\mu_0 = c_0^{-2}$, where c_0 is the velocity of light in a vacuum.

Methods of Expression

SUPERSCRIPTS AND SUBSCRIPTS

Some commonly used superscripts and subscripts are listed below.

Superscripts

- •   pure substance
- ⊖   standard value of a property (e.g. 1 atm, 1 mol kg $^{-1}$)
- ∞   limiting value at infinite dilution
- ✴   transition state (activated complex)

Subscripts

A	solvent
B, etc.	solute(s)
B, C, etc.	components of a mixture
i	typical ionic species
+, —	+ve or —ve ion
m	molar quantity
p, T, V, etc.	constant pressure, temperature, volume, etc.
f, e, s, t, d	fusion or formation, evaporation, sublimation, transition, and dissolution, respectively
c	critical state, critical value

TABLES AND GRAPHS

The value of a physical quantity is expressed as the product of a pure number and a unit,

e.g. $$p = 1 \cdot 013 \times 10^5 \text{ N m}^{-2}$$

which rearranges to $p/\text{N m}^{-2} = 1 \cdot 013 \times 10^5$

or $$p/10^5 \text{ N m}^{-2} = 1 \cdot 013$$

To avoid repetition of the unit symbol, it is common practice to tabulate data in the form of pure numbers. It follows that column headings should be dimensionless, e.g. $p/\text{N m}^{-2}$ and not $p(\text{N m}^{-2})$. A column heading such as $p(\text{N m}^{-2})$ implies pressure multiplied by N m^{-2}, when one really means pressure divided by N m^{-2}. As an example of this notation, the following tabulated data *(Table 1.9)* refer to the vapour pressure of acetone at various temperatures:

Table 1.9

$t/^{\circ}\text{C}$	T/K	$p/10^5 \text{ N m}^{-2}$	$10^3 \text{ K}/T$	$\log_{10} (p/\text{N m}^{-2})$
40	313	0·561	3·195	4·749
50	323	0·817	3·096	4·912
60	333	1·155	3·003	5·063
70	343	1·600	2·915	5·204

The same considerations apply to the labelling of graphs.

LOGARITHMIC RELATIONSHIPS IN PHYSICAL CHEMISTRY—STANDARD STATES

Since only pure numbers can be converted into the corresponding logarithm, it follows that in physico-chemical equations, such as

$$\Delta G^{\ominus} = -RT \ln K_p$$

and
$$E = E^{\ominus} + \frac{RT}{zF} \ln \frac{a(\text{oxidized})}{a(\text{reduced})}$$

K_p and $\dfrac{a(\text{oxidized})}{a(\text{reduced})}$ must be dimensionless.

K_p in the first of these equations should, strictly speaking, be written as $K_{p/p^{\ominus}}$, where $p^{\ominus}$ is a standard value of pressure, conventionally chosen to be 1 atmosphere. For example, consider the gas phase equilibrium,

$$N_2 + 3 H_2 = 2 NH_3$$

$K_p = \dfrac{(p_{NH_3})^2}{(p_{N_2})(p_{H_2})^3}$ and has the dimension [pressure] $^{-2}$, whereas

$$K_{p/p^{\ominus}} = \frac{(p_{NH_3}/p^{\ominus})^2}{(p_{N_2}/p^{\ominus})(p_{H_2}/p^{\ominus})^3} = \frac{(p_{NH_3}/1 \text{ atm})^2}{(p_{N_2}/1 \text{ atm})(p_{H_2}/1 \text{ atm})^3}$$

is dimensionless, but numerically equal to K_p expressed in the unit, atm $^{-2}$. $\Delta G^{\ominus}$ is the Gibbs free energy change for reaction at thermodynamic temperature, T, between N_2 at partial pressure $p^{\ominus}$ (1 atm), and H_2 at partial pressure, $p^{\ominus}$ (1 atm), to form NH_3 at partial pressure, $p^{\ominus}$ (1 atm). (Strictly speaking, fugacities rather than partial pressures should be considered for this equilibrium.)

Consider now the application of the Nernst equation to an electrode; for example, $Cu^{2+}(aq) \mid Cu(s)$. The equation takes the form,

$$E = E^{\ominus} + \frac{RT}{zF} \ln \frac{a[Cu^{2+}(aq)]}{a[Cu(s)]}$$

where a represents relative activity, i.e. activity relative to a selected standard state.

For a solid, pure solid is conventionally chosen as the standard state of unit activity. The relative activity of the Cu(s) in contact with $Cu^{2+}(aq)$ in this electrode is, therefore, equal to the dimensionless ratio of its activity to that of pure Cu(s) and will, of course, be unity in the absence of impurity.

For an involatile solute, a convenient choice of standard state refers to an ideal solution with a molality of 1 mol kg^{-1}. The relative activity (again, dimensionless) of the Cu^{2+}(aq) ions is, therefore, given by the expression,

$$a_{Cu^{2+}} = \frac{m_{Cu^{2-}}}{m^{\ominus}} \gamma_{Cu^{2+}}$$

where $m_{Cu^{2+}}$ is the molality of the Cu^{2+} ions expressed in the unit, mol kg^{-1}, $m^{\ominus} = 1$ mol kg^{-1} (by convention) and $\gamma_{Cu^{2-}}$ is the activity coefficient (dimensionless) of the Cu^{2+} ions. The relationship, $a = \dfrac{m}{m^{\ominus}} \gamma$, is frequently written in the abbreviated and more convenient, but strictly incorrect, form, $a = m\gamma$.

The values of standard electrode potentials, $E^{\ominus}$, commonly quoted relate specifically to the above choice of standard states, together with the convention that $\Delta G_f^{\ominus}(H^+) = 0$ and the sign convention of referring to reduction potentials.

EQUATIONS

The symbol used to denote a physical quantity should not imply the choice of a particular unit, as illustrated by the following pairs of *grammatically* incorrect and correct statements.

(1) The cryoscopic constant of a liquid is given by the approximate expression:

(a) *(incorrect)* $K_f = \dfrac{RT_f^2}{\Delta H_f}$, where ΔH_f is the latent heat of fusion

per kilogramme of liquid and T_f is the fusion temperature of the liquid:

(b) *(correct)* $K_t = \dfrac{RT_t^2}{\Delta H_t}$, where ΔH_t is the latent heat of fusion

of the liquid and T_t is its fusion temperature.

(2) According to the Debye–Hückel theory, the mean ionic activity coefficient of a 1–1 electrolyte in very dilute aqueous solution at 25°C is given by the expression:

(a) *(incorrect)* $\log_{10}\gamma_{\pm} = -0.51 \sqrt{m}$. The left hand side of this equation is dimensionless, whereas the right hand side has the dimension [amount of substance]$^{1/2}$ [mass]$^{-1/2}$; the equation is, therefore, dimensionally invalid.

(b) *(correct)* $\log_{10}\gamma_\pm = -A\sqrt{m}$, where $A = 0.51 \text{ mol}^{-1/2} \text{ kg}^{1/2}$, or, $\log_{10}\gamma_\pm = -0.51\sqrt{(m/\text{mol kg}^{-1})}$
or, $\log_{10}\gamma'_\pm = -0.51\sqrt{(m/m^\ominus)}$, where $m^\ominus$ is a standard value of molality equal to 1 mol kg^{-1}.

SPECIFIC QUANTITIES

The adjective *specific* denotes an extensive quantity divided by the mass of substance; for example, the *specific volume* of a substance is its volume divided by its mass. When an extensive quantity is denoted by a capital letter, the corresponding specific quantity can be denoted by the corresponding lower case letter; for example, $v = V/m$, $c_p = C_p/m$.

CONDUCTANCE AND CONDUCTIVITY

The term *conductance* is used to denote a quantity with the dimension [resistance] $^{-1}$, whilst *conductivity* denotes the ratio of current density to potential gradient (field strength), i.e. a quantity with the dimension [resistance] $^{-1}$ [length] $^{-1}$. In view of the meaning of the adjective specific (see previous note), neither *specific conductance* nor *specific conductivity* should be used to denote the latter of these quantities.

MOLAR QUANTITIES

The adjective *molar* denotes an extensive property divided by the amount of substance; for example, $H_m = H/n$. The SI unit for amount of substance is the mole, 1 mole of substance being $6.022\,5 \times 10^{23}$ specified elementary units (cf. Avogadro constant).

The subscript m, which denotes a molar quantity may be dropped when the appropriate unit is quoted, for example, $\Delta H^\ominus_{298\,K} = 5000$ J mol^{-1} unambiguously refers to a standard *molar* enthalpy change at 298 K and a subscript, m, is unnecessary.

Correct notation for thermochemical equations is, for example,

$$H_2(g) + \tfrac{1}{2}O_2(g) = H_2O(l); \quad \Delta H^\ominus_{298\,K} = -286 \text{ kJ mol}^{-1}$$

$$2\,H_2(g) + O_2(g) = 2\,H_2O(l); \quad \Delta H^\ominus_{298\,K} = -572 \text{ kJ mol}^{-1}$$

(not, $\Delta H^\ominus_{298\,K} = -286$ kJ and -572 kJ, respectively; the first reaction is between 1 mol of $H_2(g)$ and 1 mol of $\tfrac{1}{2}O_2(g)$ to give 1 mol of $H_2O(l)$, and the second reaction is between 1 mol of $2\,H_2(g)$ and 1 mol of $O_2(g)$ to give 1 mol of $2\,H_2O(l)$, i.e. the nature of the elementary units is specified in the equation).

The terms *molarity* and *molar concentration* and the symbol, M (which means 10^6) should not be used to describe concentration expressed in the unit, mol dm^{-3}. Molar is not recognized as a specially named SI unit of concentration, especially in view of the above definition. Similarly, neither molal nor the symbol m (which means metre or 10^{-3}) should be used to denote molality expressed in the unit, mol kg^{-1}.

RELATIVE ATOMIC AND MOLECULAR MASSES

Relative atomic mass ('atomic weight') and *relative molecular mass* ('molecular weight') are dimensionless quantities defined as the ratio or the average mass per atom or molecule (natural isotopic composition being assumed unless otherwise specified) divided by $\frac{1}{12}$ of the mass of an atom of the nuclide, ^{12}C,

e.g. $A_r(Cl) = 35\cdot453 \quad M_r(HCl) = 36\cdot461$

A_r and M_r must not be confused with *molar mass* (symbol, M) which the dimension [mass] [amount of substance]$^{-1}$ and is numerically equal to $(A_r$ or $M_r) \times 10^{-3}$ if expressed in the unit $kg\,mol^{-1}$. As an example of possible confusion between these terms, the equation giving the average velocity of a gas molecule is $\bar{c} = \left(\dfrac{8RT}{\pi M}\right)^{1/2}$, where M is the molar mass (not, as sometimes incorrectly stated, the relative atomic or molecular mass, which would make the equation dimensionally invalid).

EQUIVALENTS AND RELATED QUANTITIES

The term *equivalent* and related descriptions, such as *equivalent weight*, *normality* and *equivalent conductivity* are obsolete, since they are both unnecessary and sometimes ambiguous. For example, the 'equivalent weight' of potassium permanganate is $\frac{1}{5}M_r(KMnO_4)$ for a reaction in acid solution and $\frac{1}{3}M_r(KMnO_4)$ for a reaction in alkaline solution. Correct and incorrect notation is illustrated in the following examples:

(1) $M_r(H_2SO_4) = 98$,   i.e. $M_r(\frac{1}{2}H_2SO_4) = 49$
(not—the equivalent weight of sulphuric acid is 49).

(2) $c(KMnO_4) = 0.1$ mol dm^{-3}, i.e. $c(\frac{1}{5}KMnO_4) = 0.5$ mol dm^{-3}
(not—0.5 N $KMnO_4$).
   (3) $\Lambda^\infty(Na_2SO_4, aq) = 0.026\ \Omega^{-1}\ m^2\ mol^{-1}$ at 25°C,
i.e. $\Lambda^\infty(\frac{1}{2}Na_2SO_4, aq) = 0.013\ \Omega^{-1}\ m^2\ mol^{-1}$ at 25°C,
   (not—$\Lambda^\infty(Na_2SO_4, aq) = 0.013\ \Omega^{-1}\ m^2\ equiv^{-1}$ at 25°C).

WAVE NUMBERS

In spectroscopy a convenient parameter is the so-called *wave number*, which is the reciprocal of the wavelength and, therefore, has the dimension [length]$^{-1}$. This quantity should not, as is sometimes the unfortunate practice, be referred to as a frequency or an energy. For example, the *fundamental vibration wave number* for hydrogen is 4405 cm^{-1}, whereas the *fundamental vibration frequency* is $(2.9979 \times 10^{10}$ cm s$^{-1}) \times (4405$ cm$^{-1}) = 1.320 \times 10^{14}$ s^{-1}. (See also the energy calculation on page 8.)

Tables of Physical Constants, Conversion Factors and Relative Atomic Masses are given in the Appendices.

CHAPTER 2

Gases

EXAMPLE 2.1. VAN DER WAALS EQUATION

The critical temperature, T_c, and critical pressure, p_c, for methane are 191 K and 46.4×10^5 N m^{-2} respectively. Calculate the van der Waals constants and estimate the radius of a methane molecule.

The van der Waals equation takes the form

$$\left(p + \frac{a}{V^2}\right)(V - b) = RT$$

which rearranges to

$$p = \frac{RT}{V - b} - \frac{a}{V^2} \tag{1}$$

At the critical point there is an inflection in the p–V curve, i.e. $\left(\dfrac{\partial p}{\partial V}\right)_T$ and $\left(\dfrac{\partial^2 p}{\partial V^2}\right)_T$ are both zero. Therefore, differentiating equation (1),

$$\left(\frac{\partial p}{\partial V}\right)_T = -\frac{RT_c}{(V_c - b)^2} + \frac{2a}{V_c^3} = 0 \tag{2}$$

and

$$\left(\frac{\partial^2 p}{\partial V^2}\right)_T = \frac{2RT_c}{(V_c - b)^3} - \frac{6a}{V_c^4} = 0 \tag{3}$$

Combination of equations (1), (2) and (3) gives

$$V_c = 3b, \quad T_c = \frac{8a}{27Rb}, \quad p_c = \frac{a}{27b^2}$$

Therefore,

$$\frac{a}{b} = \frac{27 \times 8 \cdot 31 \times 191}{8} \quad \text{J mol}^{-1}$$

and

$$\frac{a}{b^2} = 27 \times 46 \cdot 4 \times 10^5 \text{ N m}^{-2}$$

from which $a = 0 \cdot 229 \text{ N m}^4 \text{ mol}^{-2}$

$$b' = 4 \cdot 28 \times 10^{-5} \text{ m}^3 \text{ mol}^{-1} = 42 \cdot 8 \text{ cm}^3 \text{ mol}^{-1}$$

The van der Waals constant, b, represents four times the actual volume of the methane molecules. Therefore, if r is the radius of a methane molecule,

$$4 / \tfrac{4}{3}\pi r^3 \times 6 \cdot 02 \times 10^{23} \text{ mol}^{-1} = 4 \cdot 28 \times 10^{-5} \text{ m}^3 \text{ mol}^{-1}$$

giving $r = 1 \cdot 62 \times 10^{-10} \text{ m}$

$$= 0 \cdot 162 \text{ nm}$$

EXAMPLE 2.2. ISOTHERMAL AND ADIABATIC EXPANSION

An ideal monatomic gas initially at 298 K and a pressure of 5 atm is expanded to a final pressure of 1 atm:

(*a*) isothermally and reversibly,
(*b*) isothermally against a constant pressure of 1 atm,
(*c*) adiabatically and reversibly,
(*d*) adiabatically against a constant pressure of 1 atm.

Calculate for each of these expansions:

(*i*)   the final temperature of the gas,
(*ii*) q, the heat absorbed by the gas,
(*iii*) w, the work done by the gas,
(*iv*) ΔU, the increase in the internal energy of the gas,
(*v*) ΔH, the increase in the enthalpy of the gas.

In the answers to these questions, subscripts $_1$ and $_2$ will be used to denote the initial and final states of the gas respectively, i.e. $p_1 = 5$ atm, $p_2 = 1$ atm, $T_1 = 298$ K.

(a) Reversible Isothermal Expansion

For an isothermal (constant temperature) process, there is no change in internal energy,

i.e. $\qquad T_2 = T_1 = 298 \text{ K}$ and $\Delta U = 0$

The work done by an ideal gas expanding isothermally and reversibly is given by

$$w = RT \ln \frac{p_1}{p_2}$$

$$= 8 \cdot 314 \times 298 \times 2 \cdot 303 \times \log_{10} 5 \text{ J mol}^{-1}$$

$$= 4000 \text{ J mol}^{-1}$$

Applying the first law of thermodynamics,

$$\Delta U = q - w$$

therefore, since $\qquad \Delta U = 0,$

$$q = w = 4000 \text{ J mol}^{-1}$$

By definition,

$$H = U + pV$$

and, therefore, $\qquad \Delta H = \Delta U + \Delta(pV)$

Since the temperature remains constant, $p_1 V_1 = p_2 V_2 = RT$, i.e. $\Delta(pV) = 0$, therefore, since $\Delta U = 0$,

$$\Delta H = 0$$

(b) Rapid Isothermal Expansion

Again, since the expansion is isothermal,

$$T_2 = T_1 = 298 \text{ K} \quad \text{and} \quad \Delta U = 0$$

The work done by an ideal gas expanding isothermally against a constant pressure, p_2, is given by

$$w = p_2(V_2 - V_1)$$

$$= p_2 \left(\frac{RT}{p_2} - \frac{RT}{p_1} \right)$$

$$= 8 \cdot 314 \times 298 \times \left(1 - \tfrac{1}{5}\right) \text{ J mol}^{-1}$$

$$= 1980 \text{ J mol}^{-1}$$

(i.e. significantly less than the maximum (reversible) value of w for isothermal expansion).

As for the reversible isothermal expansion,

$$q = w = 1980 \text{ J mol}^{-1}$$

and
$$\Delta H = \Delta U + \Delta(pV) = 0$$

(c) Reversible Adiabatic Expansion

An adiabatic process is, by definition, one in which no heat is absorbed or liberated,

i.e.
$$q = 0$$

For reversible adiabatic expansion and compression of an ideal gas

$$pV^\gamma = \text{constant}$$

where $\gamma = C_p/C_V$, C_p and C_V being the molar heat capacities at constant pressure and constant volume respectively. By the equipartition of energy principle (see Example 2.3) C_V for a monatomic gas is expressed entirely in terms of translational kinetic energy changes and is equal to $\frac{3}{2}R$ (independent of temperature over the range pertinent to this problem). Therefore, $C_p = C_V + R = \frac{5}{2}R$, and $\gamma = \frac{5}{3}$.

$$p_1 V_1^\gamma = p_2 V_2^\gamma$$

therefore, substituting $\dfrac{RT}{p}$ for V,

$$p_1^{1-\gamma} T_1^\gamma = p_2^{1-\gamma} T_2^\gamma$$

or
$$T_2 = T_1 \left(\frac{p_1}{p_2}\right)^{\frac{1-\gamma}{\gamma}}$$

$$= 298 \times (5)^{-0 \cdot 4} \text{ K}$$

$$= 157 \text{ K}$$

$$\Delta U = C_V (T_2 - T_1)$$

$$= \frac{3}{2} \times 8 \cdot 314 \times (157 - 298) \text{ J mol}^{-1}$$

$$= -1760 \text{ J mol}^{-1}$$

Applying the first law of thermodynamics,

$$\Delta U = q - w$$

therefore, since $q = 0$

$$w = -\Delta U = 1760 \text{ J mol}^{-1}$$
$$\Delta H = C_p(T_2 - T_1)$$
$$= \tfrac{5}{2} \times 8 \cdot 314 \times (157 - 298) \text{ J mol}^{-1}$$
$$= -2930 \text{ J mol}^{-1}$$

(d) Rapid Adiabatic Expansion

Again, since the expansion is adiabatic,

$$q = 0$$

and

$$\Delta U = -w$$

For expansion against a constant pressure, p_2,

$$w = p_2(V_2 - V_1)$$

Therefore, $$\Delta U = C_V(T_2 - T_1) = -p_2(V_2 - V_1)$$

Substituting $\dfrac{RT}{p}$ for V,

$$\frac{3}{2} R(T_2 - T_1) = -p_2 \left(\frac{RT_2}{p_2} - \frac{RT_1}{p_1} \right)$$

i.e. $$\frac{3}{2}(T_2 - 298 \text{ K}) = -T_2 + \frac{298 \text{ K}}{5}$$

giving $$T_2 = 203 \text{ K}$$

(i.e. the cooling is less than for the reversible adiabatic expansion).

$$\Delta U = C_V(T_2 - T_1)$$
$$= \tfrac{3}{2} \times 8 \cdot 314 \times (203 - 298) \text{ J mol}^{-1}$$
$$= -1180 \text{ J mol}^{-1}$$

and $$w = -\Delta U = 1180 \text{ J mol}^{-1}$$
$$\Delta H = C_p(T_2 - T_1)$$
$$= \tfrac{5}{2} \times 8 \cdot 314 \times (203 - 298) \text{ J mol}^{-1}$$
$$= -1970 \text{ J mol}^{-1}$$

Type of expansion	T_2/K	q/J mol^{-1}	w/J mol^{-1}	ΔU/J mol^{-1}	ΔH/J mol^{-1}
Reversible isothermal	298	4000	4000	0	0
Rapid isothermal	298	1980	1980	0	0
Reversible adiabatic	157	0	1760	−1760	−2930
Rapid adiabatic	203	0	1180	−1180	−1970

EXAMPLE 2.3. HEAT CAPACITY—EQUIPARTITION OF ENERGY PRINCIPLE

The heat capacity at constant (atmospheric) pressure of carbon dioxide is given by the expression,

$C_p = (26 \cdot 8 + 42 \cdot 3 \times 10^{-3}\, T - 14 \cdot 3 \times 10^{-6}\, T^2)$ J K^{-1} mol^{-1} where T is the thermodynamic temperature.

(a) Calculate C_p at 300 K and compare it with the value obtained according to the principle of equipartition of energy.

(b) Calculate the heat required to raise the temperature of 1 mol of CO_2 from 300 K to 700 K (i) at constant (atmospheric) pressure, and (ii) at constant volume.

(a) At $T = 300$ K

$$C_p = \{26 \cdot 8 + (42 \cdot 3 \times 10^{-3} \times 300) - [14 \cdot 3 \times 10^{-6}$$
$$\times (300)^2]\}\ \text{J K}^{-1}\,\text{mol}^{-1}$$
$$= 38 \cdot 2\ \text{J K}^{-1}\,\text{mol}^{-1}.$$

According to the equipartition principle, an energy of $\frac{1}{2}RT$ per mole is associated with each term used to describe the translational, rotational and vibrational energy of the molecules. Translational energy is described by three $\frac{1}{2}m\bar{c^2}$ terms (x, y and z directions). The rotational energy of polyatomic molecules is described by two $\frac{1}{2}I\omega^2$ terms if linear and three $\frac{1}{2}I\omega^2$ terms if non-linear. Polyatomic molecules have $3n - 5$ modes of vibration if linear and $3n - 6$ modes of vibration if non-linear, where n is the atomicity; and each mode of vibration is associated with a kinetic energy term and a potential energy term.

Carbon dioxide is a linear, triatomic molecule, therefore,

$$U - U_0 = 3 \times (\tfrac{1}{2}RT) + 2 \times (\tfrac{1}{2}RT) + 2 \times (9-5) \times (\tfrac{1}{2}RT)$$
$$\underset{\text{(trans)}}{} \qquad \underset{\text{(rot)}}{} \qquad\qquad \underset{\text{(vib)}}{}$$
$$= 6\tfrac{1}{2}RT$$

i.e. $$C_V = \left(\frac{\partial U}{\partial T}\right)_V = 6\tfrac{1}{2}R$$

and $$C_p \approx C_V + R = 7\tfrac{1}{2}R = 62\cdot4 \text{ J K}^{-1}\text{ mol}^{-1}$$

This is notably greater than the experimental value of C_p. The discrepancy arises because the vibrational contribution to C_p is much less than that predicted by the equipartition principle owing to quantization restrictions.

(*b*) $$\Delta H = \int_{T_1}^{T_2} C_p \, dT$$

$$= \int_{300 \text{ K}}^{700 \text{ K}} [26\cdot8 + (42\cdot3 \times 10^{-3}T) - (14\cdot3 \times 10^{-6}T^2)] \, dT$$
$$\text{J mol}^{-1}$$

$$= \Big(26\cdot8[700 - 300] + \frac{42\cdot3 \times 10^{-3}}{2}[(700)^2 - (300)^2]$$

$$- \frac{14\cdot3 \times 10^{-6}}{3}[(700)^3 - (300)^3]\Big) \text{ J mol}^{-1}$$

$$= (10\,720 + 8460 - 1510) \text{ J mol}^{-1}$$

$$= 17\,670 \text{ J mol}^{-1}$$

$$\Delta U \approx \Delta H - R \, \Delta T$$

$$= \{17\,670 - [8\cdot314 \times (700 - 300)]\} \text{ J mol}^{-1}$$

$$= 14\,340 \text{ J mol}^{-1}$$

EXAMPLE 2.4. MOLECULAR VELOCITIES—MAXWELL DISTRIBUTION
EQUATION

Calculate (a) the root mean square velocity, (b) the mean velocity, and (c) the most probable velocity of hydrogen molecules at 25°C.
Approximately what fraction of these molecules have a velocity in excess of the most probable velocity?

(a) Root mean square velocity, as the name implies, is defined by the expression

$$\sqrt{(\overline{c^2})} = \sqrt{\left(\frac{\sum n_i c_i^2}{\sum n_i}\right)}$$

where n_i is the number of molecules having a velocity c_i.
For an ideal gas, the kinetic theory leads to the equation

$$pV = \tfrac{1}{3}nm\overline{c^2}$$

where n is the number of molecules and m the mass of each molecule. For 1 mol of gas, $nm = M$ (the molar mass) and $pV = RT$, therefore,

$$\sqrt{(\overline{c^2})} = \sqrt{\left(\frac{3RT}{M}\right)}$$

and, since $M = 2.016\times10^{-3}$ kg mol^{-1} and $T = 298$ K,

$$\sqrt{(\overline{c^2})} = \sqrt{\left(\frac{3\times8.31\times298}{2.016\times10^{-3}}\right)} \text{ m s}^{-1}$$

$$= 1920 \text{ m s}^{-1}$$

(b) The average velocity, $\overline{c}$, is given by

$$\overline{c} = \sqrt{\left(\frac{8RT}{\pi M}\right)} = \sqrt{\left(\frac{8\times8.31\times298}{\pi\times2.016\times10^{-3}}\right)} \text{ m s}^{-1}$$

$$= 1770 \text{ m s}^{-1}$$

(c) The most probable velocity, α, is given by

$$\alpha = \sqrt{\left(\frac{2RT}{M}\right)} = \sqrt{\left(\frac{2\times8.31\times298}{2.016\times10^{-3}}\right)} \text{ m s}^{-1}$$

$$= 1570 \text{ m s}^{-1}$$

(Note that $\sqrt{(\overline{c^2})} > \overline{c} > \alpha$)

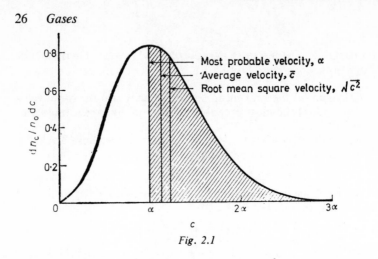

Fig. 2.1

For a total of n_0 molecules, the fraction, $\dfrac{\mathrm{d}n_c}{n_0}$, which have a velocity between c and $\mathrm{d}c$ is given by the Maxwell velocity distribution equation:

$$\frac{\mathrm{d}n_c}{n_0} = 4\pi \left(\frac{M}{2\pi RT}\right)^{3/2} c^2 \exp\left(-\frac{Mc^2}{2RT}\right)\mathrm{d}c$$

which combined with $\alpha = \sqrt{\left(\dfrac{2RT}{M}\right)}$ gives

$$\frac{\mathrm{d}n_c}{n_0} = \frac{4}{\alpha^3 \sqrt{\pi}} c^2 \exp\left(-c^2/\alpha^2\right)\mathrm{d}c$$

The fraction of the molecules with a velocity in excess of the most probable velocity can be estimated by measuring areas under the velocity distribution curve (*Fig. 2.1*), and is equal to

$$\frac{\text{shaded area}}{\text{total area}} = 0.57$$

(independent of temperature and the nature of the gas).

EXAMPLE 2.5. VISCOSITY AND MOLECULAR COLLISIONS

The coefficient of viscosity of nitrogen gas at 0°C and atmospheric pressure is 16.6×10^{-6} kg m^{-1} s^{-1}. Estimate (*a*) the collision diameter of a nitrogen molecule, and (*b*) the collision rate and mean free path for nitrogen at 0°C and a pressure of 1 atmosphere.

Assuming that the nitrogen molecules behave as perfectly elastic spheres, the coefficient of viscosity, η, is given by Chapman's equation:

$$\eta = 0.499\,\varrho\bar{c}\lambda$$

where ϱ is the density of the gas, $\bar{c}$ is the average velocity of the molecules and λ is the mean free path of the molecules. Since the gas is at s.t.p.,

$$\varrho = \frac{28\times10^{-3}}{0.0224}\ \text{kg m}^{-3}$$

and

$$\bar{c} = \sqrt{\left(\frac{8RT}{\pi M}\right)} = \sqrt{\left(\frac{8\times8.31\times273}{\pi\times28\times10^{-3}}\right)}\ \text{m s}^{-1}$$

Therefore,

$$\lambda = \frac{\eta}{0.499\varrho\bar{c}}$$

$$= \frac{16.6\times10^{-6}\times0.0224}{0.499\times28\times10^{-3}\sqrt{\left(\dfrac{8\times8.31\times273}{\pi\times28\times10^{-3}}\right)}}\ \text{m}$$

$$= 5.86\times10^{-8}\ \text{m}$$

Mean free path, λ, and collision diameter, σ, are related by the expression

$$\lambda = \frac{1}{\sqrt{2}\pi n\sigma^2}$$

where n is the concentration of molecules, i.e.

$$n = \frac{\text{Avogadro's constant}}{\text{molar volume}} = \frac{6.02\times10^{23}}{0.0224}\ \text{m}^{-3}$$

Therefore,

$$\sigma^2 = \frac{0.0224}{\sqrt{2}\times\pi\times5.86\times10^{-8}\times6.02\times10^{23}}\ \text{m}^2$$

giving

$$\sigma = 3.78\times10^{-10}\ \text{m}$$

$$= 0.378\ \text{nm}$$

The collision rate (i.e. the number of collisions taking place in unit volume of the gas in unit time) is given by the expression:

$$Z = \frac{\pi n^2\sigma^2\bar{c}}{\sqrt{2}}$$

Substituting $\sqrt{\left(\dfrac{8RT}{\pi M}\right)}$ for $\bar{c}$,

$$Z = 2n^2\sigma^2 \sqrt{\left(\dfrac{\pi RT}{M}\right)}$$

$$= 2\times\left(\dfrac{6\cdot02\times10^{23}}{0\cdot0224}\right)^2 \times(3\cdot78\times10^{-10})^2$$

$$\times\sqrt{\left(\dfrac{\pi\times8\cdot31\times273}{28\times10^{-3}}\right)} \; m^{-3}\,s^{-1}$$

$$= 1\cdot04\times10^{35}\; m^{-3}\,s^{-1}$$

Additional Examples

1. Two bulbs, 1 dm³ and 4 dm³ in volume, connected by a stopcock, are filled respectively with nitrogen at $1\cdot6\times10^5$ N m⁻² pressure and oxygen at $0\cdot6\times10^5$ N m⁻² pressure. The stopcock is opened and the gases mix. Assuming ideal behaviour, calculate (*a*) the total pressure, (*b*) the partial pressure of each gas, and (*c*) the mole fraction of each gas in the resulting mixture.

2. Nitrogen has a critical temperature of 126 K and a critical pressure of $33\cdot9\times10^5$ N m⁻². Estimate the pressure exerted by 1 mol of nitrogen at 273 K occupying a volume of 1 dm³.

3. Estimate the molar volume of ammonia at s.t.p., given that the critical temperature is 405·5 K and the critical pressure is 113×10^5 N m⁻².

4. For helium at 273 K, $pV = 100$ J when $p = 10^5$ N m⁻² and $pV = 102\cdot85$ J when $p = 27\cdot0\times10^5$ N m⁻². Neglecting the effects of intermolecular attraction, calculate (*a*) the number of moles of helium involved, and (*b*) the radius of the helium molecule.

(G.R.I.C. Part 1, 1967)

5. The density of ammonia at 298 K and at various pressures is given in the following table:

$p/10^5$ N m⁻²	0·25	0·50	0·75	1·00
$\rho/$kg m⁻³	0·172 2	0·345 3	0·519 4	0·694 3

Calculate the relative molecular mass of ammonia.

6. Calculate the maximum and minimum work required to compress 1 mol of ideal gas from 1 atm to 5 atm at a constant temperature of 298 K.

7. 10 g of CaC_2 is reacted with an excess of water at 293 K and atmospheric pressure. Calculate the work done in expansion against the atmosphere by the acetylene evolved during the reaction.

8. To what pressure must oxygen at 273 K and 10^5 N m^{-2} pressure be adiabatically and reversibly compressed in order to raise its temperature to 373 K? The mean value of C_p for oxygen over this temperature range is 29·1 J K^{-1} mol^{-1}.

9. An ideal gas, originally at 273 K and a pressure of 10^5 N m^{-2}, is compressed adiabatically to a pressure of 10^6 N m^{-2}. Taking C_V to be 21 J K^{-1} mol^{-1} and independent of temperature, calculate the maximum temperature which is attainable as a result of this compression.

10. Use the virial equation, $pV = RT + bp$ (in which $b = 26·7$ cm^3 mol^{-1}) to calculate the work done when 1 mol of hydrogen at 25°C expands isothermally from 10 atm to 1 atm against a constant pressure of 1 atm. Compare your answer with the corresponding work of expansion for 1 mol of ideal gas.

11. Calculate the heat required to raise the temperature of 1 mol of nitrogen from 0°C to 1000°C at constant (atmospheric) pressure, given that

$$C_p = (26·98 + 5·92 \times 10^{-3} T - 0·339 \times 10^{-6} T^2) \text{ J K}^{-1} \text{mol}^{-1}$$

where T is the thermodynamic temperature.

12. Calculate, according to the Equipartition of Energy Principle, the maximum molar heat capacity at constant volume at 0°C for (a) argon, (b) oxygen, and (c) hydrogen sulphide (nonlinear).

13. Calculate the temperature at which the root mean square velocity of the molecules in methane gas (assumed ideal) is 10^3 m s^{-1}.

14. Calculate, approximately, the fraction of nitrogen molecules at 25°C which have speeds between 600 m s^{-1} and 700 m s^{-1}.

15. Assuming that an evacuated system contains residual oxygen at a pressure of 10^{-6} torr (mmHg) and at 300 K, calculate (a) the number of molecules per cm^3, (b) the mean free path of the oxygen molecules, and (c) the mean time taken for an oxygen molecule to traverse its mean free path.

The collision diameter of the oxygen molecule can be taken as $3·6 \times 10^{-10}$ m.

(G.R.I.C., Part 1, 1968)

16. Calculate the collision rate and the mean free path for nitrogen at s.t.p. The collision diameter may be taken as 3.75×10^{-10} m, and the average velocity of a gas molecule in space is equal to $\sqrt{\left(\dfrac{8RT}{\pi M}\right)}$ where M represents molar mass.

Under similar conditions of temperature and pressure, how would you expect (a) average molecular velocities between collisions, and (b) mean free paths in the vapour or gaseous state to compare with those in the liquid state?

(Liverpool Polytechnic, H.N.C., 1968

17. Given the following information, calculate (a) the collision diameter of an argon molecule, and (b) the mean free path for argon at s.t.p.

Coefficient of viscosity, η, of argon at s.t.p. $= 14.58 \times 10^{-6}$ kg m^{-1} s^{-1}. $\eta = 0.499 \varrho \bar{c} \lambda$, where ϱ represents density, $\bar{c}$ average velocity and λ mean free path. $\bar{c} = \sqrt{\left(\dfrac{8RT}{\pi M}\right)}$

From the above relationships, show how you would expect the viscosity of a gas to depend on temperature and pressure.

18. The viscosity of gaseous Cl_2 at 1 atm pressure and 20°C is 1.47×10^{-4} poise (g cm^{-1} s^{-1}). Find the collision diameter of the chlorine molecule. The mean free path is given by $\lambda = (\sqrt{2}\pi n \sigma^2)^{-1}$ and the average speed of a gas molecule in space is $\bar{c} = (8RT/\pi M)^{1/2}$.

[University of Leeds, B.Sc. (3CH), 1968]

19. 100 cm^3 of oxygen effuses through a pin-hole orifice in 40 s. Under the same conditions of temperature and pressure, 100 cm^3 of a mixture of carbon monoxide and carbon dioxide effuses in 45 s. What is the composition of this mixture?

20. The Knudsen equation for the rate of bombardment of a surface by molecules is

$$z = \frac{n\bar{c}}{4}$$

where n is the number of molecules per unit volume and $\bar{c}$ is the average velocity.

Calculate the number of molecules and the number of grammes of carbon dioxide per hour which strike a leaf of 100 cm^2 surface area exposed to the atmosphere, in which the partial pressure of CO_2 is 29.1 N m^{-2} and the temperature of which is 20°C.

(G.R.I.C. Part 1, 1966)

CHAPTER 3

Thermochemistry

EXAMPLE 3.1. HESS'S LAW OF CONSTANT HEAT SUMMATION

Calculate the heat of formation of propane gas from its elements
(a) at constant pressure, (b) at constant volume, given that at
298 K and 1 atm pressure:

Heat of combustion of propane	$= -2220$ kJ mol^{-1}
Heat of formation of water	$= -286{\cdot}0$ kJ mol^{-1}
Heat of formation of carbon dioxide	$= -393{\cdot}5$ kJ mol^{-1}

(Liverpool Polytechnic, H.N.C., 1966)

Hess's law of constant heat summation is an application of the
first law of thermodynamics and states that the overall heat
change for a given process is independent of the route by which
the process takes place. Consequently, thermochemical equations
can be added to and subtracted from one another.

$$C_3H_8(g) + 5\,O_2(g) = 3\,CO_2(g) + 4\,H_2O(l);$$
$$\Delta H^{\ominus}_{298\ K} = -2220 \text{ kJ mol}^{-1} \qquad (1)$$

$$H_2(g) + \tfrac{1}{2}O_2(g) = H_2O(l); \quad \Delta H^{\ominus}_{298\ K} = -286{\cdot}0 \text{ kJ mol}^{-1} \qquad (2)$$

$$C(s) \ + \ O_2(g) = CO_2(g); \quad \Delta H^{\ominus}_{298\ K} = -393{\cdot}5 \text{ kJ mol}^{-1} \qquad (3)$$

Adding $4\times(2)$ to $3\times(3)$

$$3\,C(s) + 4\,H_2(g) + 5\,O_2(g) = 3\,CO_2(g) + 4\,H_2O(l);$$
$$\Delta H^{\ominus}_{298\ K} = -2324{\cdot}5 \text{ kJ mol}^{-1} \qquad (4)$$

Subtracting (1) from (4)

$$3\,C(s) + 4\,H_2(g) = C_3H_8(g); \quad \Delta H^{\ominus}_{298\,K} = -104{\cdot}5 \text{ kJ mol}^{-1}$$

The heat of formation of propane gas from its elements at a constant pressure of 1 atm and at 298 K, i.e. the standard heat of formation of propane at 298 K is, therefore, $-104{\cdot}5$ kJ mol^{-1}, the negative sign indicating that heat is evolved.

For a constant volume process the measured heat change represents a change, ΔU, in internal energy and is given by

$$\Delta H = \Delta U + p\,\Delta V$$

where $p\,\Delta V$ is the work of expansion when the process takes place at constant pressure. Only gases need be considered in calculating $p\,\Delta V$. Assuming ideal behaviour, $p\,\Delta V = \Delta nRT$, and

$$\Delta H = \Delta U + \Delta nRT$$

where Δn is the increase in the amount of gas. For the reaction in question, 4 mol of $H_2(g)$ are consumed for every 1 mol of $C_3H_8(g)$ produced, i.e. $\Delta n = -3$ mol mol$^{-1} = -3$, therefore,

$$-104{\cdot}5 = \Delta U_{298\,K}/\text{kJ mol}^{-1} + \frac{(-3) \times 8{\cdot}314 \times 298}{1000}$$

giving $\qquad \Delta U_{298\,K} = -97{\cdot}0 \text{ kJ mol}^{-1}$

i.e. the heat of formation of propane gas from its elements at constant volume and at 298 K is $-97{\cdot}0$ kJ mol^{-1}.

EXAMPLE 3.2. HESS'S LAW OF CONSTANT HEAT SUMMATION

Calculate the enthalpy of formation of aqueous hydrogen peroxide, given that:

$$SnCl_2(aq) + 2\,HCl(aq) + H_2O_2(aq) = SnCl_4(aq) + 2\,H_2O;$$
$$\Delta H = -371{\cdot}6 \text{ kJ mol}^{-1} \qquad (1)$$

$$SnCl_2(aq) + HCl(aq) + HOCl(aq) = SnCl_4(aq) + H_2O;$$
$$\Delta H = -314{\cdot}0 \text{ kJ mol}^{-1} \qquad (2)$$

$$2\,HI(aq) + HOCl(aq) = I_2(s) + HCl(aq) + H_2O;$$
$$\Delta H = -215{\cdot}2 \text{ kJ mol}^{-1} \qquad (3)$$

$$\tfrac{1}{2}H_2(g) + \tfrac{1}{2}I_2(s) + aq = HI(aq);$$
$$\Delta H = -\ 55{\cdot}1 \text{ kJ mol}^{-1} \qquad (4)$$

$$H_2(g) + \tfrac{1}{2}O_2(g) = H_2O(l); \qquad \Delta H = -286{\cdot}1 \text{ kJ mol}^{-1} \qquad (5)$$

[University of Durham, B.Sc. (Service 1),1967]

This is an application of Hess's law which involves reactions in aqueous solution. The symbol 'aq' denotes the presence of sufficient water for the solution in question to have negligible heat of dilution.

Subtracting $2\times(4)$ from $2\times(5)$

$$H_2(g)+O_2(g)+2\,HI(aq) = I_2(s)+2\,H_2O;$$
$$\Delta H = -462\cdot0\text{ kJ mol}^{-1} \qquad (6)$$

Subtracting (3) from (6)

$$H_2(g)+O_2(g)+HCl(aq) = HOCl(aq)+H_2O;$$
$$\Delta H = -246\cdot8\text{ kJ mol}^{-1} \qquad (7)$$

Adding (2) to (7)

$$H_2(g)+O_2(g)+SnCl_2(aq)+2\,HCl(aq) = SnCl_4(aq)+2\,H_2O;$$
$$\Delta H = -560\cdot8\text{ kJ mol}^{-1} \qquad (8)$$

Subtracting (1) from (8)

$$H_2(g)+O_2(g)+aq = H_2O_2(aq); \quad \Delta H = -189\cdot2\text{ kJ mol}^{-1}$$

i.e. the heat of formation of aqueous hydrogen peroxide at the temperature in question is $-189\cdot2$ kJ mol^{-1}.

EXAMPLE 3.3. KIRCHHOFF EQUATION

At 25°C the latent heat of sublimation of iodine is 62·3 kJ mol^{-1} and the standard enthalpy of formation of HI(g) is 24·7 kJ mol^{-1}. Calculate the enthalpy change which occurs when HI(g) is formed from the gaseous elements at 225°C. Mean molar heat capacities over the temperature range 25°C to 225°C are:

$$H_2(g); \qquad C_p = 29\cdot08\text{ J K}^{-1}\text{ mol}^{-1}$$
$$I_2(g); \qquad C_p = 33\cdot56\text{ J K}^{-1}\text{ mol}^{-1}$$
$$HI(g); \qquad C_p = 29\cdot87\text{ J K}^{-1}\text{ mol}^{-1}$$

(G.R.I.C. Part 1, 1968)

$$\tfrac{1}{2}H_2(g)+\tfrac{1}{2}I_2(s) = HI(g); \qquad \Delta H_{298\text{ K}}^{\ominus} = +24\cdot7\text{ kJ mol}^{-1} \qquad (1)$$
$$I_2(s) = I_2(g); \qquad \Delta H_{298\text{ K}} = +62\cdot3\text{ kJ mol}^{-1} \qquad (2)$$

therefore, halving (2) and subtracting from (1)

$$\tfrac{1}{2}H_2(g)+\tfrac{1}{2}I_2(g) = HI(g); \qquad \Delta H_{298\text{ K}} = -6\cdot45\text{ kJ mol}^{-1}$$

According to the Kirchhoff equation,

$$\left[\frac{\partial(\Delta H)}{\partial T}\right]_p = \Delta C_p$$

therefore,

$$\Delta H_{498\ K} - \Delta H_{298\ K} = \int_{298\ K}^{498\ K} \Delta C_p\ dT$$

where

$$\Delta C_p = \sum C_p(\text{products}) - \sum C_p(\text{reactants})$$
$$= C_p(\text{HI}) - [C_p(\tfrac{1}{2}\text{H}_2) + C_p(\tfrac{1}{2}\text{I}_2)]$$

In this problem the mean values of C_p between 298 K and 498 K are taken and the Kirchhoff equation simplifies to

$$\Delta H_{498\ K} - \Delta H_{298\ K} = \overline{\Delta C_p}(498\ K - 298\ K)$$

therefore,

$$\Delta H_{498\ K}/\text{J mol}^{-1} = -6450 + \left[29{\cdot}87 - \left(\frac{29{\cdot}08}{2} + \frac{33{\cdot}56}{2}\right)\right]$$
$$\times (498 - 298)$$

from which,

$$\Delta H_{498\ K} = -6740\ \text{J mol}^{-1}$$

i.e. $\tfrac{1}{2}\text{H}_2(\text{g}) + \tfrac{1}{2}\text{I}_2(\text{g}) = \text{HI}(\text{g}); \qquad \Delta H_{498\ K} = -6{\cdot}74\ \text{kJ mol}^{-1}$

EXAMPLE 3.4. KIRCHHOFF EQUATION

Calculate the standard enthalpy change at 473 K for the reaction, $CO + \tfrac{1}{2}O_2 = CO_2$. The standard heats of formation 'of CO and CO_2 at 298 K are $-110{\cdot}5$ kJ mol^{-1} and $-393{\cdot}5$ kJ mol^{-1}, respectively. The heat capacities of CO, CO_2 and O_2 over the temperature range 298 K to 473 K are given by the equations:

$$C_p(\text{CO}) = (26{\cdot}53 + 7{\cdot}70 \times 10^{-3}\,T - 1{\cdot}17 \times 10^{-6}\,T^2)$$
$$\text{J K}^{-1}\text{ mol}^{-1}$$

$$C_p(\text{CO}_2) = (26{\cdot}78 + 42{\cdot}26 \times 10^{-3}\,T - 14{\cdot}23 \times 10^{-6}\,T^2)$$
$$\text{J K}^{-1}\text{ mol}^{-1}$$

$$C_p(\text{O}_2) = (25{\cdot}52 + 13{\cdot}60 \times 10^{-3}\,T - 4{\cdot}27 \times 10^{-6}\,T^2)$$
$$\text{J K}^{-1}\text{ mol}^{-1}$$

where T is the thermodynamic temperature.

$$\text{C(s)} + \tfrac{1}{2}\text{O}_2(\text{g}) = \text{CO(g)}; \qquad \Delta H_{298\ K}^{\ominus} = -110{\cdot}5\ \text{kJ mol}^{-1} \qquad (1)$$
$$\text{C(s)} + \text{O}_2(\text{g}) = \text{CO}_2(\text{g}); \qquad \Delta H_{298\ K}^{\ominus} = -393{\cdot}5\ \text{kJ mol}^{-1} \qquad (2)$$

Subtracting (1) from (2),

$$CO(g) + \tfrac{1}{2}O_2(g) = CO_2(g); \quad \Delta H^{\ominus}_{298 \text{ K}} = -283.0 \text{ kJ mol}^{-1}$$

According to the Kirchhoff equation,

$$\Delta H^{\ominus}_{473 \text{ K}} - \Delta H^{\ominus}_{298 \text{ K}} = \int_{298 \text{ K}}^{473 \text{ K}} \Delta C_p \, dT$$

where $\quad \Delta C_p = C_p(CO_2) - [C_p(CO) + C_p(\tfrac{1}{2}O_2)]$

i.e.

$$\Delta C_p / \text{J K}^{-1} \text{ mol}^{-1} = \left[26.78 - \left(26.53 + \frac{25.52}{2} \right) \right] + \left\{ \left[42.26 - \left(7.70 \right. \right. \right.$$
$$\left. \left. \left. + \frac{13.60}{2} \right) \right] \times 10^{-3} T \right\} - \left\{ \left[14.23 - \left(1.17 + \frac{4.27}{2} \right) \right] \times 10^{-6} T^2 \right\}$$
$$= -12.51 + 27.76 \times 10^{-3} T - 10.92 \times 10^{-6} T^2$$

Therefore, $\quad \Delta H^{\ominus}_{473 \text{ K}} / \text{J mol}^{-1} = -283\,000 + \int_{298 \text{ K}}^{473 \text{ K}} (-12.51 + 27.76$

$$\times 10^{-3} T - 10.92 \times 10^{-6} T^2) \, dT$$
$$= -283\,000 - 12.51 \times (473 - 298) + \frac{27.76 \times 10^{-3}}{2} \times (473^2 - 298^2)$$
$$- \frac{10.92 \times 10^{-6}}{3} \times (473^3 - 298^3)$$
$$= -283\,000 - 2190 + 1870 - 290$$
$$= -283\,610$$

i.e. $\quad CO(g) + \tfrac{1}{2}O_2(g) = CO_2(g); \quad \Delta H^{\ominus}_{473 \text{ K}} = -283.6 \text{ kJ mol}^{-1}.$

EXAMPLE 3.5. ENTHALPY OF SUBLIMATION

From the following data, calculate the standard enthalpy of sublimation of ice at $-50°C$.

Mean heat capacity of ice $= 1.975 \text{ J K}^{-1} \text{ g}^{-1}$

Mean heat capacity of liquid water $= 4.185 \text{ J K}^{-1} \text{ g}^{-1}$

Mean heat capacity of water vapour $= 1.860 \text{ J K}^{-1} \text{ g}^{-1}$

Enthalpy of fusion of ice at $0°C = 333.5 \text{ J g}^{-1}$

Enthalpy of evaporation of water at $100°C = 2255.2 \text{ J g}^{-1}$.

(G.R.I.C. Part 1, 1968)

$\Delta H_{223\ K}^{\ominus}$ (sublimation of ice) $= \Delta H_{223\ K}^{\ominus}$ (fusion of ice)$+\Delta H_{223\ K}^{\ominus}$ (evaporation of water)

Applying Hess's law:

$\Delta H_{223\ K}^{\ominus}$ (fusion of ice) $= \Delta H^{\ominus}$ (ice at 223 K to ice at 273 K)

$+\Delta H_{273\ K}^{\ominus}$ (fusion of ice)$+\Delta H^{\ominus}$ (water at 273 K to water at 223 K)

$$= (1{\cdot}975\times 50+333{\cdot}5-4{\cdot}185\times 50)\ \text{J g}^{-1}$$
$$= 223\ \text{J g}^{-1}$$

Similarly

$\Delta H_{223\ K}^{\ominus}$ (evaporation of water) $= \Delta H^{\ominus}$(water at 223 K to

water at 373 K)$+\Delta H_{373\ K}^{\ominus}$ (evaporation of water)

$+\Delta H^{\ominus}$ (water vapour at 373 K to water vapour at 223 K)

$$= (4{\cdot}185\times 150+2255{\cdot}2-1{\cdot}860\times 150)\ \text{J g}^{-1}$$
$$= 2604\ \text{J g}^{-1}$$

Therefore

$$\Delta H_{223\ K}^{\ominus}\text{ (sublimation of ice)} = (223+2604)\ \text{J g}^{-1}$$
$$= 2827\ \text{J g}^{-1}$$

Note. In this calculation, the Kirchhoff equation has, in effect, been used to calculate $\Delta H_{223\ K}^{\ominus}$ (fusion of ice) and $\Delta H_{223\ K}^{\ominus}$ (evaporation of water).

EXAMPLE 3.6. BOND ENERGY

Calculate the average O—H bond energy in water from the following data:

$H_2O(l) = H_2O(g)$;	$\Delta H = +\ 40{\cdot}6\ \text{kJ mol}^{-1}$	(1)
$2\ H(g) = H_2(g)$;	$\Delta H = -435{\cdot}0\ \text{kJ mol}^{-1}$	(2)
$O_2(g) = 2\ O(g)$;	$\Delta H = +489{\cdot}6\ \text{kJ mol}^{-1}$	(3)
$2\ H_2(g)+O_2(g) = 2\ H_2O(l)$;	$\Delta H = -571{\cdot}6\ \text{kJ mol}^{-1}$	(4)

(University of Sheffield, B.Sc. 1st year, 1967)

Bond energy is the average energy required to break a particular bond in a molecule to form atoms and/or radicals. In the water molecule there are two O—H bonds and the average O—H bond energy is half of the heat change in the process

$$H_2O(g) = 2 H(g) + O(g)$$

Subtracting $(1) + (2) + \frac{1}{2} \times (4)$ from $\frac{1}{2} \times (3)$ gives

$$H_2O(g) = 2 H(g) + O(g), \quad \Delta H = 925 \text{ kJ mol}^{-1}$$

The average O—H bond energy in water is, therefore, $462 \cdot 5 \text{ kJ mol}^{-1}$.

EXAMPLE 3.7. RESONANCE ENERGY

From the following thermochemical data, calculate the resonance energy of toluene, for which the Kekulé formula is

—CH₃

$$(\Delta H^{\ominus}_{298 \text{ K}}/\text{kJ mol}^{-1})$$

$C_7H_8(l) + 9 O_2(g)$	$= 7 CO_2(g) + 4 H_2O(l),$	$- 3910$	(1)
$C_7H_8(l)$	$= C_7H_8(g),$	$38 \cdot 1$	(2)
$H_2(g)$	$= 2 H(g),$	$436 \cdot 0$	(3)
$C(s)$	$= C(g),$	$715 \cdot 0$	(4)
$H_2(g) + \frac{1}{2}O_2(g)$	$= H_2O(l),$	$-285 \cdot 8$	(5)
$C(s) + O_2(g)$	$= CO_2(g),$	$-393 \cdot 5$	(6)

Bond energies/kJ mol⁻¹:

$$E(\text{C—H}) = 413 \cdot 0, \quad E(\text{C—C}) = 345 \cdot 6, \quad E(\text{C}=\text{C}) = 610 \cdot 0$$

(University of Nottingham, B.Sc. 1st year, 1966)

The energy of dissociation of gaseous toluene into carbon atoms and hydrogen atoms is calculated by applying Hess's law to the above thermochemical equations as follows:

$$(1) + 4 \times (3) + 7 \times (4) - (2) - 4 \times (5) - 7 \times (6) \quad \text{gives}$$

$$C_7H_8(g) = 7 C(g) + 8 H(g), \quad \Delta H^{\ominus}_{298 \text{ K}} = 6698 \cdot 6 \text{ kJ mol}^{-1}$$

The Kekulé formula for toluene includes 8 (C—H) bonds, 4 (C—C) bonds and 3 (C=C) bonds. The energy of dissociation of gaseous toluene calculated from bond energies is, therefore,

$$[(8 \times 413 \cdot 0) + (4 \times 345 \cdot 6) + (3 \times 610 \cdot 0)] \text{ kJ mol}^{-1} = 6516 \cdot 4 \text{ kJ mol}^{-1}.$$

The difference between these two values for the energy of dissociation, i.e. $182 \cdot 2 \text{ kJ mol}^{-1}$, is the resonance energy of toluene.

EXAMPLE 3.8. BORN–HABER CYCLE

Calculate the lattice energy of KCl(s) from the following data:

Standard enthalpy of formation of potassium chloride	$= -435$ kJ mol^{-1}
Standard enthalpy of sublimation of metallic potassium	$= 88$ kJ mol^{-1}
Ionization potential of potassium	$= 410$ kJ mol^{-1}
Electron affinity of chlorine atoms	$= 368$ kJ mol^{-1}
Standard enthalpy of dissociation of chlorine	$= 226$ kJ mol^{-1}

The lattice energy, ΔH_c, is the enthalpy change for the process,

$$KCl(s) \rightarrow K^+(g) + Cl^-(g)$$

and can be calculated from the data given by consideration of the following cyclic process (Born–Haber cycle):

$$KCl(s) \xrightarrow{\Delta H_c} K^+(g) + Cl^-(g)$$

$$\downarrow -\Delta H_f \qquad -e \Big| I \qquad +e \Big| -A$$

$$K(s) + \tfrac{1}{2}Cl_2(g) \xrightarrow{\Delta H_s + \frac{1}{2}\Delta H_d} K(g) + Cl(g)$$

where ΔH_f is the enthalpy of formation of KCl(s), ΔH_s is the enthalpy of sublimation of potassium, ΔH_d is the enthalpy of dissociation of $Cl_2(g)$, I is the ionization potential of potassium and A is the electron affinity of chlorine atoms ($K - e \rightarrow K^+$ is endothermic and $Cl + e \rightarrow Cl^-$ is exothermic).

Applying Hess's law to this cyclic process,

$$\Delta H_c = -\Delta H_f + \Delta H_s + \tfrac{1}{2}\Delta H_d + I - A$$

$$= \left(435 + 88 + \frac{226}{2} + 410 - 368\right) \text{ kJ mol}^{-1}$$

$$= 678 \text{ kJ mol}^{-1}$$

Additional Examples

1. Given the following heats of combustion:

H_2;	$\Delta H^{\ominus}_{298 \text{ K}} = -286 \cdot 0$ kJ mol^{-1}
C;	$\Delta H^{\ominus}_{298 \text{ K}} = -393 \cdot 5$ kJ mol^{-1}
C_2H_2;	$\Delta H^{\ominus}_{298 \text{ K}} = -1301$ kJ mol^{-1}
C_2H_4;	$\Delta H^{\ominus}_{298 \text{ K}} = -1410$ kJ mol^{-1}
C_2H_6;	$\Delta H^{\ominus}_{298 \text{ K}} = -1561$ kJ mol^{-1}

calculate $\Delta H^{\ominus}_{298 \text{ K}}$ for the formation of acetylene, ethylene and ethane, and $\Delta H^{\ominus}_{298 \text{ K}}$ for the hydrogenation of acetylene and ethylene to ethane.

2. Calculate ΔH for the reaction, $C_2H_2(g) + H_2O(l) = CH_3CHO(l)$, given that:

$$2 C(s) + H_2(g) = C_2H_2(g), \quad \Delta H = + 226 \cdot 8 \text{ kJ mol}^{-1}$$

$$H_2(g) + \tfrac{1}{2}O_2(g) = H_2O(l), \quad \Delta H = - 286 \cdot 0 \text{ kJ mol}^{-1}$$

$$C(s) + O_2(g) = CO_2(g), \quad \Delta H = - 393 \cdot 5 \text{ kJ mol}^{-1}$$

$$CH_3CHO(l) + 2\tfrac{1}{2} O_2(g) = 2 CO_2(g) + 2 H_2O(l),$$

$$\Delta H = -1167 \quad \text{kJ mol}^{-1}$$

3. The standard heats of formation at 298 K of toluene, carbon dioxide and water are $+48 \cdot 0 \text{ kJ mol}^{-1}$, $-393 \cdot 5 \text{ kJ mol}^{-1}$ and $-286 \cdot 0 \text{ kJ mol}^{-1}$, respectively. Calculate the heat evolved when 10 g of liquid toluene is completely combusted at 298 K and at constant (atmospheric) pressure.

4. The heats of formation of CO and CO_2 at constant pressure and 298 K are $-110 \cdot 5 \text{ kJ mol}^{-1}$ and $-393 \cdot 5 \text{ kJ mol}^{-1}$, respectively. Calculate the corresponding heats of formation at constant volume.

5. The increase in temperature (at around 25°C) arising from the combustion of 1 g of benzoic acid in a bomb calorimeter can be duplicated by the application of 26·4 kJ of electrical energy. Given that the standard heats of formation of water and carbon dioxide are $-286 \cdot 0 \text{ kJ mol}^{-1}$ and $-393 \cdot 5 \text{ kJ mol}^{-1}$, respectively, at 25°C, calculate the standard heat of formation of benzoic acid at 25°C.

6. Calculate $\Delta H^{\ominus}$ for the reaction

$$FeO(s) + 2 H^+(aq) = H_2O(l) + Fe^{2+}(aq)$$

given that

$$2 Fe(s) + \tfrac{3}{2}O_2(g) = Fe_2O_3(s); \qquad \Delta H^{\ominus} = -822 \cdot 2 \text{ kJ mol}^{-1}$$

$$2 FeO(s) + \tfrac{1}{2}O_2(g) = Fe_2O_3(s); \qquad \Delta H^{\ominus} = -284 \cdot 1 \text{ kJ mol}^{-1}$$

$$H_2(g) + \tfrac{1}{2}O_2(g) = H_2O(l); \qquad \Delta H^{\ominus} = -286 \cdot 0 \text{ kJ mol}^{-1}$$

$$\tfrac{1}{2}H_2(g) = H^+(aq); \qquad \Delta H^{\ominus} = 0$$

$$Fe(s) + 2 H^+(aq) = Fe^{2+}(aq) + H_2(g); \quad \Delta H^{\ominus} = - 86 \cdot 2 \text{ kJ mol}^{-1}$$

[University of Durham, B.Sc. (service 1), 1968]

7. Use the following data to calculate $\Delta H^{\ominus}_{298 \text{ K}}$ for the formation of NaOH(aq)

$$\Delta H^{\ominus}_{298 \text{ K}} \text{ (ionization of water)} \quad = + \ 56 \cdot 9 \text{ kJ mol}^{-1}$$

$$\Delta H^{\ominus}_{298 \text{ K}} \text{ (formation of water)} \quad = -286 \cdot 0 \text{ kJ mol}^{-1}$$

$$\Delta H^{\ominus}_{298 \text{ K}} \text{ (formation of Na}^{+}\text{(aq))} = -240 \cdot 6 \text{ kJ mol}^{-1}$$

8. From the following thermochemical equations, calculate the heat of formation of nitrous oxide from its elements at 298 K:

$$C + N_2O \ = CO + N_2; \qquad \Delta H_{298 \text{ K}} = -192 \cdot 9 \text{ kJ mol}^{-1}$$

$$C + O_2 \quad = CO_2; \qquad \Delta H_{298 \text{ K}} = -393 \cdot 5 \text{ kJ mol}^{-1}$$

$$2 CO + O_2 = 2CO_2; \qquad \Delta H_{298 \text{ K}} = -566 \cdot 0 \text{ kJ mol}^{-1}$$

The heat capacities of nitrogen, oxygen and nitrous oxide in the temperature range 298 K to 423 K are given by the following equations:

$$C_p(N_2) \ = (27 \cdot 0 + 0 \cdot 006T) \text{ J K}^{-1} \text{ mol}^{-1}$$

$$C_p(O_2) \ = (25 \cdot 6 + 0 \cdot 014T) \text{ J K}^{-1} \text{ mol}^{-1}$$

$$C_p(N_2O) = (27 \cdot 2 + 0 \cdot 044T) \text{ J K}^{-1} \text{ mol}^{-1}$$

where T is the thermodynamic temperature. Calculate the heat of formation of nitrous oxide at 423 K.

9. The enthalpy of formation of ammonia gas at 298 K is $-46 \cdot 2 \text{ kJ mol}^{-1}$. The mean molar heat capacities of gaseous H_2, N_2 and NH_3 in the temperature range 250 K to 450 K are given by the following equations:

$$C_p(H_2) \ = (29 \cdot 1 + 0 \cdot 002T) \text{ J K}^{-1} \text{ mol}^{-1}$$

$$C_p(N_2) \ = (27 \cdot 0 + 0 \cdot 006T) \text{ J K}^{-1} \text{ mol}^{-1}$$

$$C_p(NH_3) = (25 \cdot 9 + 0 \cdot 032T) \text{ J K}^{-1} \text{ mol}^{-1}$$

where T is the thermodynamic temperature. Calculate ΔH and ΔU for the formation of ammonia at 398 K.

10. Given the following data:

$$CO(g) + H_2O(g) = CO_2(g) + H_2(g); \qquad \Delta H_{298 \text{ K}} = -42 \cdot 0 \text{ kJ mol}^{-1}$$

$$C_p(CO) \ = (26 \cdot 8 + \ 6 \times 10^{-3} T) \text{ J K}^{-1} \text{ mol}^{-1}$$

$$C_p(H_2O) = (30 \cdot 1 + 10 \times 10^{-3} T) \text{ J K}^{-1} \text{ mol}^{-1}$$

$$C_p(H_2) \ = (28 \cdot 5 + \ 2 \times 10^{-3} T) \text{ J K}^{-1} \text{ mol}^{-1}$$

$$C_p(CO_2) = (26 \cdot 4 + 42 \times 10^{-3} T) \text{ J K}^{-1} \text{ mol}^{-1}$$

where T is the thermodynamic temperature, calculate ΔH for the above reaction at 1000 K.

11. Calculate the dissociation energy of the C—Cl bond in methyl chloride from the following data:

Enthalpy of formation of $CH_3Cl(g) = -$ 82·0 kJ mol^{-1}

Enthalpy of formation of $CH_4(g)$ $= -$ 74·9 kJ mol^{-1}

Enthalpy of atomization of C $=$ 715·0 kJ mol^{-1}

Enthalpy of atomization of H_2 $=$ 436·0 kJ mol^{-1}

Enthalpy of atomization of Cl_2 $=$ 241·8 kJ mol^{-1}

12. The bond dissociation energy of the first hydrogen–sulphur bond in hydrogen sulphide, D (H—SH), is 376·6 kJ mol^{-1}. Calculate the heat of formation ($\Delta H_f^{\ominus}$) and the bond dissociation energy of the free radical, HS*, given that:

$$H_2S(g), \quad \Delta H_f^{\ominus} = - \ 20\cdot1 \ kJ\,mol^{-1}$$
$$S(g), \quad \Delta H_f^{\ominus} = \ 277\cdot0 \ kJ\,mol^{-1}$$
$$D\,(H—H) = \ 436\cdot0 \ kJ\,mol^{-1}$$

[University of Nottingham, B.Sc. (Part 1), 1966]

13. In the combustion of solid naphthalene at 298 K and atmospheric pressure, 5157 kJ mol^{-1} of heat are evolved. Use this result together with the data below (all of which are $\Delta H_{298\,K}$ terms) to calculate the resonance energy of naphthalene.

Latent heat of sublimation of naphthalene $=$ 72·0 kJ mol^{-1}

Enthalpy of formation of H_2O (l) $= -286\cdot0$ kJ mol^{-1}

Enthalpy of formation of CO_2 $= -393\cdot5$ kJ mol^{-1}

Enthalpy of atomization of H_2 $=$ 436·0 kJ mol^{-1}

Enthalpy of atomization of C $=$ 715·5 kJ mol^{-1}

Thermochemical bond energy, C—C $=$ 345·5 kJ mol^{-1}

Thermochemical bond energy, C=C $=$ 617·0 kJ mol^{-1}

Thermochemical bond energy, C—H $=$ 413·0 kJ mol^{-1}

(G.R.I.C. Part 2, 1966)

14. Calculate the strain energy of cyclopropane, C_3H_6, and cyclopentane, C_5H_{10}, using the data given. Comment on the values obtained.

$$\Delta H_f[C_3H_6(g)] = 53\cdot2 \text{ kJ mol}^{-1}$$
$$\Delta H_f[C_5H_{10}(g)] = -77\cdot2 \text{ kJ mol}^{-1}$$
$$\Delta H_f[C(g)] = 715\cdot0 \text{ kJ mol}^{-1}$$
$$\Delta H_f[H(g)] = 218\cdot0 \text{ kJ mol}^{-1}$$
$$E(C-C) = 355\cdot6 \text{ kJ mol}^{-1}$$
$$E(C-H) = 407\cdot9 \text{ kJ mol}^{-1}$$

15. Calculate the electron affinity of bromine atoms from the following data.

Lattice energy of sodium bromide $= 736 \text{ kJ mol}^{-1}$

Standard enthalpy of formation of sodium bromide $= -376 \text{ kJ mol}^{-1}$

Standard enthalpy of sublimation of sodium $= 109 \text{ kJ mol}^{-1}$

Standard enthalpy of atomization of bromine $= 192 \text{ kJ mol}^{-1}$

Ionization potential of sodium $= 490 \text{ kJ mol}^{-1}$

CHAPTER 4

Thermodynamics and Chemical Equilibria

EXAMPLE 4.1. HEAT ENGINE—REFRIGERATION

A refrigerator, which works at 50 per cent of ideal efficiency, operates with its interior at 0°C in a room at 25°C. Calculate the energy required to freeze 1 kg of water (initially at 0°C), also the heat discharged into the room during this process. The latent heat of fusion of ice is 334 J g^{-1}.

[Liverpool Polytechnic, B.Sc. (1st year), 1968]

A heat engine operates between an upper (source) temperature, T_2, and a lower (sink) temperature, T_1. The engine absorbs a quantity of heat, q_2, at the upper temperature. This is partially converted into mechanical work, w, and the remainder, q_1, is liberated as heat at the lower temperature,

i.e. $$w = q_2 - q_1$$

The maximum efficiency, η_{max}, is given by

$$\eta_{max} = \frac{w}{q_2} = \frac{T_2 - T_1}{T_2}$$

A refrigerator is a heat engine operating in reverse. Work, w, is done on the engine, causing heat q_1 to be absorbed at the lower temperature, T_1, and heat q_2 (where $q_2 = q_1 + w$), is liberated a the higher temperature. Using the above expressions, the maximum efficiency, η_{max}, may, in this case, be expressed as

$$\eta_{max} = \frac{q_1}{w} = \frac{T_1}{T_2 - T_1}$$

and, if the refrigerator is operating at 50 per cent of its ideal efficiency,

$$\frac{q_1}{w} = \frac{0 \cdot 5 T_1}{T_2 - T_1}$$

If 1 kg of water (initially at 0°C) is frozen in the refrigerator,

$$q_1 = 1000 \times 334 \text{ J} = 334\,000 \text{ J}$$

Therefore, $$\frac{334\,000 \text{ J}}{w} = \frac{0 \cdot 5 \times 273}{298 - 273}$$

giving w (the required energy input) $= 61\,200$ J
The heat discharged into the room at 25°C is given by

$$q_2 = q_1 + w$$
$$= 334\,000 \text{ J} + 61\,200 \text{ J}$$
$$= 395\,200 \text{ J}$$

EXAMPLE 4.2. ENTROPY CHANGE ON MIXING

Calculate the change in entropy when 10 g of ice at 0°C is added to 50 g of water at 40°C in an isolated system. The latent heat of fusion of ice is 334·4 J g^{-1} and the specific heat of water is 4·18 J K^{-1} g^{-1}.

Let θ°C be the equilibrium temperature after mixing the ice and water. The heat absorbed when 10 g of ice at 0°C is converted to water at θ°C equals

$$(10 \times 334 \cdot 4) \text{ J} + (10 \times 4 \cdot 18 \times \theta) \text{ J}$$

and the heat evolved when 50 g of water is cooled from 40°C to θ°C equals

$$50 \times 4 \cdot 18 \times (40 - \theta) \text{ J}$$

Equating these heat changes gives $\theta = 20$°C.

The increase in entropy, ΔS_1, when 10 g of ice at 0°C is converted to water at 20°C is given by

$$\Delta S_1 = \frac{\Delta H_f}{T_f} + \int_{273\,\text{K}}^{293\,\text{K}} \frac{C_p}{T}\,\mathrm{d}T$$

where T represents thermodynamic temperature, T_f is the melting point of ice (273 K), ΔH_f is the heat of fusion of the ice and C_p is the heat capacity of the water,

i.e. $\Delta S_1 = \dfrac{10\times334\cdot4}{273} \text{ J K}^{-1} + 10\times4\cdot18\times2\cdot303\times\log_{10}\dfrac{293}{273} \text{ J K}^{-1}$

$\qquad = 15\cdot21 \text{ J K}^{-1}$

The decrease in entropy, ΔS_2, on cooling 50 g of water from 40°C to 20°C is given by

$$\Delta S_2 = \int_{293\,\text{K}}^{313\,\text{K}} \frac{C_p}{T}\,\mathrm{d}T$$

$\qquad = 50\times4\cdot18\times2\cdot303\times\log_{10}\dfrac{313}{293} \text{ J K}^{-1}$

$\qquad = 13\cdot77 \text{ J K}^{-1}$

The net entropy change is, therefore,

$\Delta S = \Delta S_1 - \Delta S_2$

$\qquad = 15\cdot21 \text{ J K}^{-1} - 13\cdot77 \text{ J K}^{-1} = +1\cdot44 \text{ J K}^{-1}$

Note. Since the mixing is a thermodynamically irreversible process, ΔS must necessarily be positive.

EXAMPLE 4.3. ENTROPY CHANGES IN AN IDEAL GAS

Calculate the entropy change when 2 mol of nitrogen (assumed to behave as an ideal gas) initially at 300 K and at atmospheric pressure are heated to 600 K (*a*) at constant pressure, and (*b*) at constant volume. C_p for nitrogen in the temperature range 300 K to 600 K, is given by the equation,

$$C_p = (27\cdot0 + 0\cdot0060\,T) \text{ J K}^{-1}\,\text{mol}^{-1}$$

where T is the thermodynamic temperature.

(a) At constant pressure, the entropy change per mole of ideal gas is given by

$$\Delta S = \int_{T_1}^{T_2} \frac{C_p}{T} \, dT$$

Therefore, for 2 mol of nitrogen heated from 300 K to 600 K

$$\Delta S = 2 \times \int_{300\,K}^{600\,K} \left(\frac{27 \cdot 0}{T} + 0 \cdot 006\,0 \right) dT \quad J\,K^{-1}$$

$$= 2 \times \left[27 \cdot 0 \times 2 \cdot 303 \times \log_{10} \frac{600}{300} + 0 \cdot 006\,0 \times (600 - 300) \right] J\,K^{-1}$$

$$= 41 \cdot 0 \ J\,K^{-1}$$

(b) When an ideal gas is also subjected to a pressure change (from an initial pressure p_1 at temperature T_1, to a final pressure p_2 at temperature T_2) the entropy change per mole is given by

$$\Delta S = \int_{T_1}^{T_2} \frac{C_p}{T} \, dT + R \ln \frac{p_1}{p_2}$$

Since constant volume is maintained, $\dfrac{p_1}{p_2} = \dfrac{T_1}{T_2} = \dfrac{300}{600}$, therefore, for 2 mol of nitrogen,

$$\Delta S = 2 \times \int_{300\,K}^{600\,K} \left(\frac{27 \cdot 0}{T} + 0 \cdot 006\,0 \right) dT \quad J\,K^{-1}$$

$$+ 2 \times 8 \cdot 314 \times 2 \cdot 303 \times \log_{10} \frac{300}{600} \quad J\,K^{-1}$$

$$= 41 \cdot 0 \ J\,K^{-1} - 11 \cdot 5 \ J\,K^{-1}$$

$$= 29 \cdot 5 \ J\,K^{-1}$$

EXAMPLE 4.4. ENTROPY AND FREE ENERGY OF MIXING

Calculate the changes in entropy and free energy which result when 4 mol of nitrogen and 1 mol of oxygen (assumed to behave as ideal gases) are mixed at 298 K and constant pressure.

For an ideal solution or a mixture of ideal gases, the increase in entropy which results when the components are mixed at constant temperature and pressure is given by the expression,

$$\Delta S_{\text{mix}} = -R \sum n_i \ln x_i$$

where n_i is the amount and x_i the mole fraction of component i.

$$n(N_2) = 4 \text{ mol}, \quad x(N_2) = 0.8$$
$$n(O_2) = 1 \text{ mol}, \quad x(O_2) = 0.2$$

Therefore,

$$\Delta S_{\text{mix}} = -8.314 \times 2.303 \times (4 \log_{10} 0.8 + \log_{10} 0.2) \text{ J K}^{-1}$$
$$= +20.8 \text{ J K}^{-1}$$
$$\Delta G_{\text{mix}} = \Delta H_{\text{mix}} - T \Delta S_{\text{mix}}$$

therefore, since $\Delta H_{\text{mix}} = 0$,

$$\Delta G_{\text{mix}} = -298 \times 20.8 \text{ J}$$
$$= -6200 \text{ J}$$

EXAMPLE 4.5. ISOTHERMAL EXPANSION

An ideal gas initially at 298 K and a pressure of 5 atm is expanded to a final pressure of 1 atm (*a*) isothermally and reversibly, (*b*) isothermally against a constant pressure of 1 atm. Calculate for each of these expansions (*i*) ΔS, the increase in the entropy of the gas, (*ii*) ΔA, the increase in the Helmholtz free energy of the gas, (*iii*) ΔG, the increase in the Gibbs free energy of the gas.

This example is a continuation of parts (*a*) and (*b*) of Example 2.2, the answers to which are requoted in the following table:

Type of expansion	T_2/K	$q/\text{J mol}^{-1}$	$w/\text{J mol}^{-1}$	$\Delta U/\text{J mol}^{-1}$	$\Delta H/\text{J mol}^{-1}$
Reversible isothermal	298	4000	4000	0	0
Rapid isothermal	298	1980	1980	0	0

Subscripts 1 and 2 refer to the initial and final states of the gas, respectively, i.e. $p_1 = 5$ atm, $p_2 = 1$ atm and $T_1 = 298$ K.

(*a*) *Reversible Isothermal Expansion*

For a reversible change in an ideal gas,

$$\Delta S = \int_{T_1}^{T_2} \frac{C_V}{T}\, dT + \int_{V_1}^{V_2} \frac{R}{V}\, dV$$

or

$$\Delta S = \int_{T_1}^{T_2} \frac{C_p}{T}\, dT - \int_{p_1}^{p_2} \frac{R}{p}\, dp$$

Since T is constant, then (using the second of these equations),

$$\Delta S = 0 - \left(2 \cdot 303 \times 8 \cdot 314 \times \log_{10} \tfrac{1}{5}\right) \text{ J K}^{-1}\text{mol}^{-1}$$
$$= +13 \cdot 4 \text{ J K}^{-1}\text{mol}^{-1}$$

$$\Delta A = \Delta U - T\ \Delta S$$
$$= 0 - (298 \times 13 \cdot 4) \text{ J mol}^{-1}$$
$$= -4000 \text{ J mol}^{-1}$$

and

$$\Delta G = \Delta H - T\ \Delta S$$
$$= 0 - (298 \times 13 \cdot 4) \text{ J mol}^{-1}$$
$$= -4000 \text{ J mol}^{-1}$$

i.e.

$$-\Delta A = -\Delta G = w_{\text{max (reversible)}}$$

(*b*) *Rapid Isothermal Expansion*

Since S, A and G are state functions, ΔS, ΔA and ΔG depend only on the initial and final states of the gas. T_1, T_2, p_1 and p_2 are the same for both the reversible and the irreversible isothermal expansion, therefore, ΔS, ΔA and ΔG must also be the same, i.e.

$$\Delta S = +13 \cdot 4 \text{ J K}^{-1}\text{mol}^{-1}$$
$$\Delta A = -4000 \text{ J mol}^{-1}$$
$$\Delta G = -4000 \text{ J mol}^{-1}$$

and

$$-\Delta A = -\Delta G > w_{\text{irreversible}}$$

EXAMPLE 4.6. FREE ENERGY CHANGE

Calculate the standard free energy change at 25°C for the reaction

$$H_2(g) + CO_2(g) = H_2O(l) + CO(g)$$

given the following standard heats of formation and entropies:

Substance	$H_2(g)$	$CO_2(g)$	$H_2O(l)$	$CO(g)$
$\Delta H_{f, 298 K}^{\ominus}$/kJ mol^{-1}	0	-393.5	-241.8	-110.5
$S_{298 K}^{\ominus}$/J K^{-1} mol^{-1}	130.5	213.8	188.7	197.9

For the reaction

$$H_2(g) + CO_2(g) = H_2O(l) + CO(g)$$

the standard enthalpy change at 298 K is given from the standard enthalpies of formation at 298 K by

$$\Delta H_{298 K}^{\ominus} = \Delta H_{f, 298 K}^{\ominus}[H_2O(l)] + \Delta H_{f, 298 K}^{\ominus}[CO(g)]$$
$$- \Delta H_{f, 298 K}^{\ominus}[CO_2(g)]$$
$$= [-241.8 + (-110.5) - (-393.5)] \text{ kJ mol}^{-1}$$
$$= 41.2 \text{ kJ mol}^{-1}$$

The standard entropy change at 298 K is given by

$$\Delta S_{298 K}^{\ominus} = S_{298 K}^{\ominus}[H_2O(l)] + S_{298 K}^{\ominus}[CO(g)] - S_{298 K}^{\ominus}[H_2(g)]$$
$$- S_{298 K}^{\ominus}[CO_2(g)]$$
$$= [188.7 + 197.9 - 130.5 - 213.8] \text{ J K}^{-1} \text{mol}^{-1}$$
$$= 42.3 \text{ J K}^{-1} \text{mol}^{-1}$$

From the relationship

$$\Delta G = \Delta H - T \Delta S$$

the standard free energy change for the reaction is given by

$$\Delta G_{298 K}^{\ominus} = [41\,200 - 298 \times 42.3] \text{ J mol}^{-1}$$
$$= 28.6 \text{ kJ mol}^{-1}$$

EXAMPLE 4.7. REACTION ISOTHERM

At 300 K and 1 atm pressure, dinitrogen tetroxide is 20 per cent dissociated into nitrogen dioxide. Calculate the equilibrium constant and standard free energy change for the reaction

$$N_2O_4 = 2\,NO_2$$

at this temperature.

The dissociation of dinitrogen tetroxide and the equilibrium pressures can be represented as follows, where p is the total pressure and α is the degree of dissociation:

$$N_2O_4 \quad = \quad 2\,NO_2$$

$$\frac{(1-\alpha)p}{1+\alpha} \qquad \frac{2\alpha p}{1+\alpha}$$

The equilibrium constant, K_p, is, therefore, given by

$$K_p = \frac{(p_{NO_2})^2}{p_{N_2O_4}}$$

$$= \frac{\left(\dfrac{2\alpha p}{1+\alpha}\right)^2}{\dfrac{(1-\alpha)p}{1+\alpha}} = \frac{4\alpha^2 p}{(1-\alpha)(1+\alpha)}$$

For 20 per cent dissociation, $\alpha = 0\cdot2$, and at a total pressure of 1 atm,

$$K_p = \frac{4\times(0\cdot2)^2\times1}{(1-0\cdot2)(1+0\cdot2)} \text{ atm}$$

$$= 0\cdot167 \text{ atm}$$

The standard free energy change, $\Delta G^{\ominus}$ is related to K_p (expressed in terms of pressures relative to a standard pressure of 1 atm, see page 13) by the reaction isotherm,

$$\Delta G^{\ominus} = -RT \ln (K_p/\text{atm})$$
$$= (-2\cdot303\times8\cdot314\times300 \log_{10} 0\cdot167) \text{ J mol}^{-1}$$
$$= 4\cdot47 \text{ kJ mol}^{-1}$$

EXAMPLE 4.8. FREE ENERGY CHANGE AND CHEMICAL AFFINITY

For the equilibrium

$$CO(g) + H_2O(g) = CO_2(g) + H_2(g)$$

the standard enthalpy and entropy changes at 300 K and 1200 K for the forward reaction are as follows:

$$\Delta H^{\ominus}_{300\ K} = -41 \cdot 16\ kJ\ mol^{-1} \qquad \Delta H^{\ominus}_{1200\ K} = -32 \cdot 93\ kJ\ mol^{-1}$$
$$\Delta S^{\ominus}_{300\ K} = -42 \cdot 4\ J\ K^{-1}\ mol^{-1} \qquad \Delta S^{\ominus}_{1200\ K} = -29 \cdot 6\ J\ K^{-1}\ mol^{-1}$$

In which direction will the reaction be spontaneous (*a*) at 300 K and (*b*) at 1200 K, when $p_{CO} = p_{H_2O} = p_{CO_2} = p_{H_2} = 1$ atm? Calculate K_p for the reaction at each temperature.

Using the relationship, $\Delta G = \Delta H - T\,\Delta S$, then, for the reaction,

$$CO(g) + H_2O(g) = CO_2(g) + H_2(g)$$

$$
\begin{aligned}
\Delta G^{\ominus}_{300\ K} &= \Delta H^{\ominus}_{300\ K} - 300\ K \times \Delta S^{\ominus}_{300\ K} \\
&= -41\ 160\ J\ mol^{-1} - 300 \times (-42 \cdot 4)\ J\ mol^{-1} \\
&= -28\ 440\ J\ mol^{-1}
\end{aligned}
$$

and
$$
\begin{aligned}
\Delta G^{\ominus}_{1200\ K} &= \Delta H^{\ominus}_{1200\ K} - 1200\ K \times \Delta S^{\ominus}_{1200\ K} \\
&= -32\ 930\ J\ mol^{-1} - 1200 \times (-29 \cdot 6)\ J\ mol^{-1} \\
&= +2590\ J\ mol^{-1}
\end{aligned}
$$

A spontaneous process involves a decrease in free energy, therefore, the reaction,

$$CO(g, 1\ atm) + H_2O(g, 1\ atm) \rightarrow CO_2(g, 1\ atm) + H_2(g, 1\ atm),$$

is spontaneous at 300 K, whereas the reverse reaction is spontaneous at 1200 K.

Equilibrium constant and standard free energy change are related by the reaction isotherm, $\Delta G^{\ominus} = -RT\ \ln K_p$. Therefore,

$$\log_{10} K_p(300\ K) = \frac{-(-28\ 400)}{2 \cdot 303 \times 8 \cdot 314 \times 300}$$

giving
$$K_p(300\ K) = 8 \cdot 8 \times 10^4$$

and
$$\log_{10} K_p(1200\ K) = \frac{-2590}{2 \cdot 303 \times 8 \cdot 314 \times 1200}$$

giving
$$K_p(1200\ K) = 0 \cdot 77$$

EXAMPLE 4.9. VAN'T HOFF ISOCHORE

The following table gives the equilibrium constant at different temperatures for the reaction,

$$2 NO(g) + O_2(g) = 2 NO_2(g)$$

T/K	600	700	800	900	1000
K_p/atm^{-1}	140	5·14	0·437	0·062 5	0·013 1

Calculate the mean standard enthalpy change for the reaction in this temperature range.

The standard enthalpy change, $\Delta H^\ominus$, is calculated using the so-called van't Hoff isochore,

$$\frac{d \ln K_p}{dT} = \frac{\Delta H^\ominus}{RT^2}$$

which on integration takes the form,

$$\log_{10} K_p = \frac{-\Delta H^\ominus}{2 \cdot 303RT} + \text{constant}$$

If the temperature-dependence of $\Delta H^\ominus$ is small, a plot of $\log_{10} K_p$ against $1/T$ should approximate to a straight line of slope $\dfrac{-\Delta H^\ominus}{2 \cdot 303R}$.

T/K	600	700	800	900	1000
$K \times 10^3/T$	1·667	1·429	1·250	1·111	1·000
$\log_{10}(K_p/\text{atm}^{-1})$	2·146	0·711	$\bar{1}$·641	$\bar{2}$·796	$\bar{2}$·117

From the graph (*Fig. 4.1*),

$$\frac{-\Delta H^\ominus/\text{J mol}^{-1}}{2 \cdot 303 \times 8 \cdot 314} = 6 \cdot 04 \times 10^3$$

giving $\qquad \Delta H^\ominus = -116 \times 10^3 \text{ J mol}^{-1}$

$$= -116 \text{ kJ mol}^{-1}$$

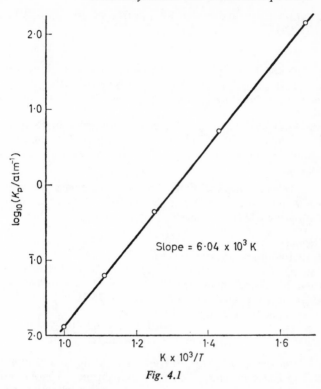

Fig. 4.1

Additional Examples

1. What is the maximum efficiency of a heat engine operating with steam under pressure between 413 K and 313 K. What minimum amount of heat must be withdrawn from the hot reservoir to obtain 8 kJ of work? How much heat is liberated at the lower temperature?

Given a hot reservoir of 473 K, what must the temperature of the cold reservoir be for the engine to operate with the same efficiency as above?

2. Calculate the change in entropy when 100 g of ice at 223 K is added to 100 g of water at 323 K in an isolated system. The latent heat of fusion of water is 334 J g^{-1}, the specific heat of water is 4·2 J K^{-1} g^{-1} and that of ice is 2·0 J K^{-1} g^{-1}.

3. Calculate the entropy change when pure nitrogen, hydrogen and ammonia gases all at s.t.p. are mixed to form 1 mol of mixture

at s.t.p. having a final molar composition of 20% N_2, 50% H_2, and 30% NH_3.

[University of Salford, B.Sc. (Chem. Eng., 2nd year), 1967]

4. Calculate the change in enthalpy and the change in entropy when 1 mol of hydrogen sulphide is (a) heated from 373 K to 473 K at a pressure of 1 atm, (b) compressed from 1 atm to 10 atm at 298 K.

The molar heat capacity of gaseous hydrogen sulphide is given by $C_p = (36 \cdot 86 + 0 \cdot 007\ 9\ T)$ J K^{-1} mol^{-1}, where T is the thermodynamic temperature. Assume that hydrogen sulphide behaves as an ideal gas.

5. From the following data for water, calculate the entropy change when 1 mol of water is heated from $-50°C$ to $500°C$ at constant atmospheric pressure.

Enthalpy of fusion of ice at 0°C	$= 6004$ J mol^{-1}
Enthalpy of vaporization of water at 100°C	$= 40\ 660$ J mol^{-1}
Mean heat capacity of ice	$= 35 \cdot 56$ J K^{-1} mol^{-1}
Mean heat capacity of liquid water	$= 75 \cdot 31$ J K^{-1} mol^{-1}
Heat capacity of steam at constant pressure	$= (30 \cdot 20 + 0 \cdot 009\ 92\ T)$ J K^{-1} mol^{-1}

where T is the thermodynamic temperature.

6. The latent heat of vaporization of water is 40·0 kJ mol^{-1} at 373 K and 1 atm pressure. For the vaporization of 1 mol of water under these conditions, calculate the external work done and the changes in internal energy (U), enthalpy (H), Gibbs free energy (G), entropy (S) and Helmholtz free energy (A).

(G.R.I.C. Part 1, 1967)

7. Use the thermodynamic data listed below to calculate $\Delta H_{298\ K}^{\ominus}$ and $\Delta G_{298\ K}^{\ominus}$ for the reactions,

$$2\,C + 2\,H_2 = C_2H_4 \tag{1}$$

$$2\,C + 3\,H_2 = C_2H_6 \tag{2}$$

and discuss the feasibility of synthesizing ethylene and ethane by direct reaction between carbon and hydrogen.

Substance	$S_{298\ K}^{\ominus}$/J K^{-1} mol^{-1}	$\Delta H_{combustion,\ 298\ K}^{\ominus}$/kJ mol^{-1}
C	5·9	-394
H_2	130·5	-286
C_2H_4	218·8	-1393
C_2H_6	230·1	-1560

8. Using the data tabulated below, discuss the thermodynamic feasibility of synthesizing ethylene at 298 K, (*a*) by reacting acetylene with hydrogen, and (*b*) by conversion of ethane to ethylene and hydrogen. Discuss in each case whether a change in temperature or pressure would be advantageous.

	$\Delta G_{f, 298\ K}^{\ominus}/\mathrm{kJ\ mol^{-1}}$	$\Delta H_{f, 298\ K}^{\ominus}/\mathrm{kJ\ mol^{-1}}$
$C_2H_6(g)$	$-32\cdot9$	$-84\cdot7$
$C_2H_2(g)$	$209\cdot2$	$226\cdot7$
$C_2H_4(g)$	$68\cdot1$	$52\cdot3$

9. At 457 K and under a total pressure of 1 atm, nitrogen dioxide is 5 per cent dissociated according to the equation

$$2\,NO_2 = 2\,NO + O_2$$

Calculate the equilibrium constants K_p and K_c for this reaction.

[University of Liverpool, B.Sc. (Part 1), 1967]

10. At 620 K and 1 atm pressure in the presence of a suitable catalyst, *n*-pentane isomerizes to form both isopentane and neopentane, i.e.

$$CH_3CH_2CH_2CH_2CH_3 = (CH_3)_2CHCH_2CH_3$$
$$CH_3CH_2CH_2CH_2CH_3 = C(CH_3)_4$$

The standard free energies of formation at 620 K of *n*-pentane, isopentane and neopentane are $141\cdot5$ kJ mol^{-1}, $136\cdot5$ kJ mol^{-1} and $146\cdot6$ kJ mol^{-1}, respectively. Calculate the composition of the mixture of isomers when equilibrium is attained.

[University of Manchester, B.Sc. (1st year), 1968]

11. When 1 mol of acetic acid is mixed with $0\cdot18$ mol of ethyl alcohol at 282 K, $0\cdot171$ mol of ester is formed at equilibrium: when 1 mol of acetic acid and 1 mol of alcohol are mixed at 282 K, $0\cdot667$ mol of ester is formed at equilibrium. Calculate the concentration equilibrium constant K_c in both cases and check that the law of mass action is obeyed. What is the standard free energy change of the esterification reaction?

[University of Bristol, B.Sc. (Subsidiary exam), 1968]

12. Under equilibrium conditions at 500 K and 1 atm total pressure, the gas nitrosyl chloride is 27 per cent dissociated according to the reaction

$$2\,NOCl = 2\,NO + Cl_2$$

Calculate the standard free energy change at this temperature. Show that for very slight dissociation, the percentage dissociation would be inversely proportional to the cube root of the equilibrium pressure. Explain briefly what information would be necessary to calculate the extent of dissociation at some other temperature.

(G.R.I.C. Part 1, 1969)

13. The equilibrium constant K_p for the dissociation of dinitrogen tetroxide into nitrogen dioxide is $1 \cdot 34$ atm at 60°C and $6 \cdot 64$ atm at 100°C. Determine the free energy change of this reaction at each temperature, and the mean heat content (enthalpy) change over the temperature range.

(G.R.I.C. Part 1, 1964)

14. Hydrogen is $0 \cdot 39$ per cent dissociated at 2200 K and $1 \cdot 61$ per cent dissociated at 2500 K, the total pressure being 1 atm in each case. Calculate K_p for the dissociation at these two temperatures and determine the mean enthalpy of dissociation over the given temperature range.

15. The equilibrium constant for the dissociation of hydrogen iodide was measured at the following temperatures:

Temperature/K	595	666	720	781	870
K_p	$0 \cdot 036 \, 0$	$0 \cdot 025 \, 2$	$0 \cdot 020 \, 9$	$0 \cdot 016 \, 8$	$0 \cdot 013 \, 3$

Calculate the mean standard enthalpy change of the reaction over this temperature range.

16. Calculate K_p and $\Delta G^{\ominus}$ for the water-gas reaction,

$$CO_2(g) + H_2(g) = CO(g) + H_2O(g)$$

at 1573 K from the information that 63 per cent of an equimolar mixture of CO_2 and H_2 is converted into CO and H_2O at equilibrium. From the free energies of formation given below, determine $\Delta G^{\ominus}_{298 \, K}$ for the reaction, and assuming that $\Delta H^{\ominus}$ and $\Delta S^{\ominus}$ are independent of temperature, evaluate the latter quantities.

	$CO_2(g)$	$H_2(g)$	$CO(g)$	$H_2O(g)$
$\Delta G^{\ominus}_{f, \, 298 \, K}$/kJ mol^{-1}	$-394 \cdot 4$	0	$-132 \cdot 3$	$-228 \cdot 6$

[University of Birmingham, B.Sc. (1st year), 1966]

CHAPTER 5

Solutions and Phase Equilibria

EXAMPLE 5.1. CLAPEYRON EQUATION

The specific volumes of water and ice at 0°C and at atmospheric pressure are 1.0001 cm³ g⁻¹ and 1.0907 cm³ g⁻¹, respectively, and the latent heat of fusion of ice is 334 J g⁻¹. Calculate the melting point of ice under a pressure of 10 MN m⁻².

For a solid–liquid phase change, the Clapeyron equation takes the form,

$$\frac{\mathrm{d}T}{\mathrm{d}p} = \frac{T_f(V_1 - V_s)}{\Delta H_f}$$

where $\dfrac{\mathrm{d}T}{\mathrm{d}p}$ is the rate of change of melting point with pressure, T_f is the melting point (in K), ΔH_f is the latent heat of fusion and V_1 and V_s are the volumes of liquid and solid, respectively.

To a close approximation, constant values of T_f, ΔH_f, V_1 and V_s can be used over the pressure range in question,

i.e. $$\frac{\Delta T}{\Delta p} = \frac{273 \times (1.000\,1 - 1.090\,7) \times 10^{-6}}{334} \quad \frac{\mathrm{K\ m^3\ g^{-1}}}{\mathrm{J\ g^{-1}}}$$

$$= -7.41 \times 10^{-8} \frac{\mathrm{K}}{\mathrm{N\ m^{-2}}}$$

At $p = 1.013 \times 10^5$ N m⁻², $T_f = 0°C$.

Substituting $\Delta p/\text{N m}^{-2} = 10^7 - (1 \cdot 013 \times 10^5) = 99 \times 10^5$,

gives, $\Delta T = -7 \cdot 41 \times 10^{-8} \times 99 \times 10^5 \text{ K}$
$$= -0 \cdot 73 \text{ K}$$

i.e. the melting point of ice under a pressure of 10 MN m^{-2} (nearly 100 atm) is $-0 \cdot 73°\text{C}$.

EXAMPLE 5.2. CLAUSIUS–CLAPEYRON EQUATION

The vapour pressure of benzene is $0 \cdot 153 \times 10^5$ N m^{-2} at 303 K and $0 \cdot 520 \times 10^5$ N m^{-2} at 333 K. Calculate the mean latent heat of evaporation of benzene over this temperature range.

The Clausius–Clapeyron equation, $\dfrac{\text{d} \ln p}{\text{d}T} = \dfrac{\Delta H_e}{RT^2}$, is used. Integrating this equation between vapour pressures p_2 and p_1, and temperatures T_2 and T_1, gives

$$2 \cdot 303 \log_{10} \frac{p_2}{p_1} = \frac{-\Delta H_e}{R} \left(\frac{1}{T_2} - \frac{1}{T_1} \right)$$

$$= \frac{+\Delta H_e}{R} \left(\frac{T_2 - T_1}{T_1 T_2} \right)$$

where ΔH_e is the mean latent heat of evaporation over the temperature range in question,

i.e. $2 \cdot 303 \log_{10} \dfrac{0 \cdot 520 \times 10^5}{0 \cdot 153 \times 10^5} = \dfrac{\Delta H_e/\text{J mol}^{-1}}{8 \cdot 314} \left(\dfrac{30}{303 \times 333} \right)$

from which $\Delta H_e = 34\ 200 \text{ J mol}^{-1}$
$$= 34 \cdot 2 \text{ kJ mol}^{-1}$$

EXAMPLE 5.3. CLAUSIUS–CLAPEYRON EQUATION

The vapour pressures of carbon tetrachloride at various temperatures are given in the following table:

Temperature/°C	20	30	40	50	60	70	80
Vapour pressure/10^5 N m^{-2}	0·121	0·191	0·288	0·423	0·601	0·829	1·124

Calculate (*a*) the mean molar heat of evaporation of carbon tetrachloride in the above temperature range, and (*b*) the molar entropy of evaporation of carbon tetrachloride at its normal boiling point.

Integrating the Clausius–Clapeyron equation, $\dfrac{\mathrm{d}\ln p}{\mathrm{d}T} = \dfrac{\Delta H_e}{RT^2}$, gives

$$\log_{10} p = \frac{-\Delta H_e}{2\cdot303RT} + \text{constant}$$

Since the temperature-dependence of the latent heat of evaporation, ΔH_e, is small, a plot of $\log_{10} p$ against $\dfrac{1}{T}$ (where p represents vapour pressure and T the thermodynamic temperature) should approximate to a straight line of slope, $\dfrac{-\Delta H_e}{2\cdot303R}$.

T/K	293	303	313	323	333	343	353
$\mathrm{K}\times10^3/T$	3·413	3·300	3·195	3·096	3·003	2·915	2·833
$\log_{10}(p/\mathrm{N\ m^{-2}})$	4·083	4·281	4·459	4·626	4·779	4·919	5·051

The slight curvature of the graph (*Fig. 5.1*) reflects the small temperature-dependence of ΔH_e. From the mean slope,

$$\frac{-(\Delta H_e/\mathrm{J\ mol^{-1}})}{2\cdot303\times8\cdot314} = -1\cdot66\times10^3$$

giving
$$\Delta H_e = 31\ 800\ \mathrm{J\ mol^{-1}}$$
$$= 31\cdot8\ \mathrm{kJ\ mol^{-1}}$$

The entropy of evaporation, $\Delta S_e^{\ominus}$, at the normal boiling point, $T_e^{\ominus}$, is given by

$$\Delta S_e^{\ominus} = \frac{\Delta H_e^{\ominus}}{T_e^{\ominus}}$$

Interpolation on the graph at $p^{\ominus} = 1\cdot013\times10^5\ \mathrm{N\ m^{-2}}$ [i.e. at $\log_{10}(p^{\ominus}/\mathrm{N\ m^{-2}}) = 5\cdot006$] gives $\dfrac{1}{T_e^{\ominus}} = 2\cdot86\times10^{-3}\ \mathrm{K^{-1}}$. The slope of the graph at this temperature is $-1\cdot57\times10^3\ \mathrm{K}$,

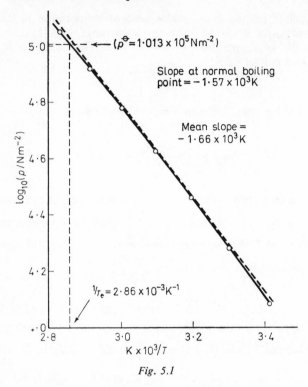

Fig. 5.1

therefore, $\Delta H_e^{\ominus} = 2\cdot303 \times 8\cdot314 \times 1\cdot57 \times 10^3$ J mol^{-1}

and, $\Delta S_e^{\ominus} = 2\cdot303 \times 8\cdot314 \times 1\cdot57 \times 10^3 \times 2\cdot86 \times 10^{-3}$ J K^{-1} mol^{-1}

$= 86$ J K^{-1} mol^{-1}

This is in accordance with Trouton's rule which states that the entropy of evaporation at the normal boiling point is approximately 85–90 J K^{-1} mol^{-1} for all normal liquids.

EXAMPLE 5.4. DISTILLATION

The saturated vapour pressures of benzene and toluene are both given by the equation,

$$\log_{10}(p^{\bullet}/\text{N m}^{-2}) = \frac{-0.052\,23\,A}{T} + B$$

where T is the thermodynamic temperature and A and B have the following values:

	A	B
Benzene	32 295 K	9·779 5
Toluene	39 198 K	10·454 9

Assuming that mixtures of benzene and toluene form ideal solutions, calculate the molar percentage of benzene in (*a*) a mixture which boils at 97°C under an external pressure of 1 atm ($1·013 \times 10^5$ N m^{-2}), and (*b*) the initial condensate formed on distilling this mixture.

(G.R.I.C. Part 1, 1967)

(*a*) For benzene (B) at 97°C (370·15 K)

$$\log_{10} (p_B^\bullet/\text{N m}^{-2}) = \frac{-0·052\ 23 \times 32\ 295}{370·15} + 9·779\ 5$$

$$= 5·222\ 5$$

giving $\qquad p_B^\bullet = 1·669 \times 10^5 \text{ N m}^{-2}$

For toluene (T) at 97°C (370·15 K)

$$\log_{10} (p_T^\bullet/\text{N m}^{-2}) = \frac{|-0·052\ 23 \times 39\ 198}{370·15} + 10·454\ 9$$

$$= 4·923\ 9$$

giving $\qquad p_T^\bullet = 0·839 \times 10^5 \text{ N m}^{-2}$

Let $x_B(l)$ be the mole fraction of benzene and, therefore, $1 - x_B(l)$ the mole fraction of toluene, in a mixture of benzene and toluene boiling at 97°C under an external pressure of $1·013 \times 10^5$ N m^{-2}. If Raoult's law is followed,

$$x_B(l) \times 1·669 \times 10^5 + [1 - x_B(l)] \times 0·839 \times 10^5 = 1·013 \times 10^5$$

giving $\qquad x_B(l) = 0·210$

i.e. the liquid mixture contains 21·0 mole per cent benzene.

(*b*) The mole fraction, $x_B(g)$, of benzene in the vapour (i.e. in the initial condensate formed on distilling the above mixture) is equal to the partial pressure of the benzene, p_B, divided by the total pressure ($1·013 \times 10^5$ N m^{-2}). The partial pressure of benzene is equal to the mole fraction of benzene in the liquid multiplied by the vapour pressure of pure benzene,

i.e.
$$p_B = x_B(l) \times p_B^{\bullet}$$
$$= 0 \cdot 210 \times 1 \cdot 669 \times 10^5 \text{ N m}^{-2}$$

and
$$x_B(g) = \frac{0 \cdot 210 \times 1 \cdot 669 \times 10^5}{1 \cdot 013 \times 10^5}$$

$$= 0 \cdot 346$$

i.e. the initial condensate contains 34·6 mole per cent benzene.

EXAMPLE 5.5. STEAM DISTILLATION

A mixture of bromobenzene and water (immiscible liquids) is distilled at standard atmospheric pressure. Given the following vapour pressures of the pure liquids, calculate (*a*) the boiling point of the mixture, and (*b*) the mass ratio of bromobenzene and water in the distillate;

Temperature/°C	92	94	96	98	100
$p(H_2O)/10^5$ N m^{-2}	0·756	0·814	0·877	0·943	1·013
$p(C_6H_5Br)/$ 10^5 N m^{-2}	0·141	0·152	0·164	0·176	0·188

Since bromobenzene and water are immiscible, the vapour pressure exerted by each liquid in the mixture is that of the pure liquid at the temperature in question. The total vapour pressure of the mixture is, therefore, given by

$$p(\text{total}) = p_{\bullet}^{\bullet}(H_2O) + p_{\bullet}^{\bullet}(C_6H_5Br)$$

The boiling point of the mixture is the temperature at which $p(\text{total})$ is equal to the external pressure ($1 \cdot 013 \times 10^5$ N m^{-2} in this case). By interpolation on the $p(\text{total})$ curve (*Fig. 5.2*), this temperature is seen to be 95·2°C.

The partial pressure of each component is proportional to the amount (number of moles) of the component present in the vapour, i.e.

$$\frac{n(C_6H_5Br)}{n(H_2O)} = \frac{p(C_6H_5Br)}{p(H_2O)}$$

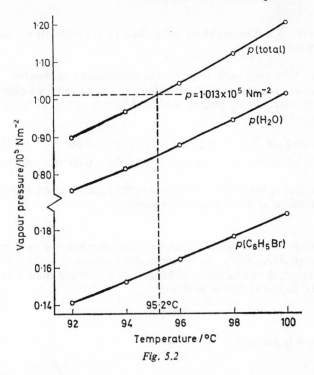

Fig. 5.2

By interpolation on the partial vapour pressure curves,

$$p(C_6H_5Br) \quad \text{at} \quad 95 \cdot 2°C = 0 \cdot 160 \times 10^5 \text{ N m}^{-2}$$

and $\quad p(H_2O) \quad \text{at} \quad 95 \cdot 2°C = 0 \cdot 853 \times 10^5 \text{ N m}^{-2}$

The mass ratio of bromobenzene to water in the distillate is, therefore, given by,

$$\frac{\text{mass of } C_6H_5Br}{\text{mass of } H_2O} = \frac{n(C_6H_5Br) \times M(C_6H_5Br)}{n(H_2O) \times M(H_2O)}$$

$$= \frac{0 \cdot 160 \times 10^5 \times 157 \times 10^{-3}}{0 \cdot 853 \times 10^5 \times 18 \times 10^{-3}}$$

$$= 1 \cdot 64$$

EXAMPLE 5.6. DISTRIBUTION OF A SOLUTE BETWEEN IMMISCIBLE SOLVENTS

The following data refer to the distribution of benzoic acid between benzene and water at 6°C, c_B and c_W being equilibrium concentrations of benzoic acid in the benzene and water layers, respectively:

$c_B/$ mol dm^{-3}	0·015 6	0·049 5	0·083 5	0·195
$c_W/$mol dm^{-3}	0·003 29	0·005 79	0·007 49	0·011 4

Neglecting the slight dissociation in the aqueous layer, show that benzoic acid is dimerized in benzene.

If benzoic acid exists as C_6H_5COOH in water and in an associated form, $(C_6H_5COOH)_n$, in benzene, then, assuming ideal solution behaviour, the distribution of benzoic acid between these solvents will be described by the equation,

$$\frac{(c_B)^{1/n}}{c_W} = k$$

where k is a constant.

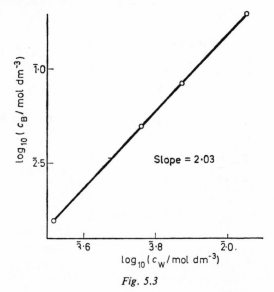

Fig. 5.3

Taking logs,

$$\frac{1}{n} \log_{10} c_{\text{B}} = \log_{10} c_{\text{w}} + \log_{10} k$$

A plot of $\log_{10} c_{\text{B}}$ against $\log_{10} c_{\text{w}}$ should, accordingly, be a straight line of slope equal to n.

$\log_{10} (c_{\text{B}}/\text{mol dm}^{-3})$ $\bar{2}\cdot 193$ $\bar{2}\cdot 695$ $\bar{2}\cdot 922$ $\bar{1}\cdot 290$

$\log_{10} (c_{\text{w}}/\text{mol dm}^{-3})$ $\bar{3}\cdot 517$ $\bar{3}\cdot 763$ $\bar{3}\cdot 875$ $\bar{2}\cdot 057$

From the graph, $n = 2\cdot03$, which is consistent with dimerization of benzoic acid in benzene.

EXAMPLE 5.7. VAPOUR PRESSURE LOWERING AND BOILING POINT ELEVATION

Given that the latent heat of evaporation of water at 100°C is 2250 J g^{-1}, calculate, assuming ideal behaviour, (a) the vapour pressure at 100°C, and (b) the boiling point, under an external pressure of $1\cdot013\times10^5$ N m^{-2}, of a solution of 50 g of glucose ($M_{\text{r}} = 180$) in 1 kg of water.

(a) If p is the vapour pressure of the solution and $p^{\bullet}$ the vapour pressure of pure solvent, then, assuming ideal behaviour and applying Raoult's law,

$$x_{\text{B}} = \frac{p^{\bullet} - p}{p^{\bullet}}$$

where x_{B} is the mole fraction of the solute.
At 100°C, $p^{\bullet} = 1\cdot013\times10^5$ N m^{-2}, therefore,

$$x_{\text{B}} = \frac{\dfrac{50}{180}}{\dfrac{1000}{18} + \dfrac{50}{180}} = 0\cdot004\,975 = \frac{1\cdot013\times10^5 - (p/\text{N m}^{-2})}{1\cdot013\times10^5}$$

from which $\qquad p = 1\cdot008\times10^5$ N m^{-2}

(b) The solution boils under an external pressure $p^{\bullet}$ (i.e. $1\cdot013\times 10^5$ N m^{-2}) when the temperature is raised from $T^{\bullet}$ (the boiling point of pure solvent) where the vapour pressure of the solution is p, to T, where the vapour pressure of the solution is $p^{\bullet}$.

Applying the Clausius–Clapeyron equation in its integrated form (see Example 5.2),

$$\ln \frac{p}{p^{\bullet}} = \frac{-\Delta H_e}{R} \left(\frac{T - T^{\bullet}}{TT^{\bullet}} \right)$$

where ΔH_e is the latent heat of evaporation of the solvent, assumed to be constant between $T^{\bullet}$ and T.

Since $p^{\bullet} - p \ll p^{\bullet}$, $\ln \dfrac{p}{p^{\bullet}}$ can be approximated as follows:

$$\ln \frac{p}{p^{\bullet}} = \ln \left[1 - \left(\frac{p^{\bullet} - p}{p^{\bullet}} \right) \right] \approx - \left(\frac{p^{\bullet} - p}{p^{\bullet}} \right)$$

Also, as an approximation, $TT^{\bullet}$ can be replaced by $(T^{\bullet})^2$. The errors introduced by this approximation and by assuming constant ΔH are somewhat self-cancelling.

Therefore,

$$\frac{p^{\bullet} - p}{p^{\bullet}} = x_B = 0.004\,975 = \frac{\Delta H_e}{R} \left(\frac{T - T^{\bullet}}{(T^{\bullet})^2} \right)$$

i.e.

$$0.004\,975 = \frac{2250 \times 18 \times (T - T^{\bullet})/\mathrm{K}}{8.314 \times (373.15)^2}$$

from which, $T - T^{\bullet} = 0.142$ K

i.e. $T = 373.29$ K

EXAMPLE 5.8. RELATIVE MOLECULAR MASS FROM FREEZING POINT DEPRESSION

Pure benzene freezes at $5.40°$C and a solution of 0.223 g of phenyl acetic acid ($C_6H_5CH_2COOH$) in 4.4 g of benzene freezes at $4.47°$C. The latent heat of fusion of benzene is 9.89 kJ mol^{-1}. Calculate the apparent relative molecular mass of phenyl acetic acid and comment on the result.

[University of Nottingham, B.Sc. (Part 1), 1965]

The freezing point depression, for an ideal dilute solution is given by the approximate expression,

$$T_f^{\bullet} - T_f = \frac{R(T_f^{\bullet})^2 n_B}{\Delta H_f (n_A + n_B)}$$

where T_f is the freezing point of the solution, $T_f^{\bullet}$ is the freezing point and ΔH_f the latent heat of fusion of the solvent, and n_A and n_B are the amounts (number of moles) of solvent and solute, respectively. Therefore, if M_r is the apparent relative molecular mass of phenyl acetic acid,

$$5\cdot40-4\cdot47 = \frac{8\cdot314\times(278\cdot5)^2\times\left(\dfrac{0\cdot223}{M_r}\right)}{9890\times\left(\dfrac{4\cdot4}{78}+\dfrac{0\cdot223}{M_r}\right)}$$

from which, $M_r = 273$

Since $M_r(C_6H_5CH_2COOH) = 136$, the above result suggests that phenyl acetic acid is dimerized in solution in benzene.

EXAMPLE 5.9. OSMOTIC PRESSURE

Calculate for an aqueous sucrose solution at 20°C with a molality equal to $1\cdot00$ mol kg^{-1}, (a) the ideal osmotic pressure (without approximation), (b) the ideal osmotic pressure according to the Morse equation, (c) the ideal osmotic pressure according to the van't Hoff equation. (At 20°C, 1 mol sucrose per kg water is equivalent to $0\cdot825$ mol sucrose per dm^3 solution and the density of water is $0\cdot998$ g cm^{-3}.)

Compare these calculated osmotic pressures with the experimental value of $27\cdot2\times10^5$ N m^{-2} and comment on the differences.

(a) The osmotic pressure, Π, of an ideal solution is given by the expression,

$$\Pi V_m = -RT \ln x_A \qquad (1)$$

where V_m is the molar volume and x_A the mole fraction of the solvent. Therefore, for $1\cdot00$ mol kg^{-1} aqueous sucrose solution at 20°C,

$$(\Pi/\text{N m}^{-2})\times\frac{18\cdot015\times10^{-3}}{0\cdot998\times10^3} = -2\cdot303\times8\cdot314\times293\cdot15$$

$$\times\log_{10}\left(\frac{\dfrac{1000}{18\cdot015}}{1+\dfrac{1000}{18\cdot015}}\right)$$

from which, $\Pi = 24\cdot0\times10^5$ N m^{-2}

(*b*) Equation (1) can be written in the form

$$\Pi V_m = -RT \ln (1 - x_B)$$

where x_B is the mole fraction of the solute. If x_B is small, then,

$$\ln (1 - x_B) \approx -x_B, \quad \text{and} \quad \Pi V_m \approx RT x_B$$

Also, if x_B is small, $x_B \approx \dfrac{n_B}{n_A}$, where n_A and n_B are the amounts (number of moles) of solvent and solute, respectively, in the solution, therefore,

$$\Pi V_m n_A \approx RT n_B$$

i.e. $$\Pi V \approx RT n_B \quad \text{(Morse equation)} \qquad (2)$$

where V is the volume of solvent in which an amount, n_B, of solute is dissolved. Therefore, for 1.00 mol kg^{-1} aqueous sucrose solution at 20°C,

$$(\Pi / \text{N m}^{-2}) \times \frac{1}{0.998 \times 10^3} \approx 8.314 \times 293 \times 1.00$$

from which, $$\Pi \approx 24.3 \times 10^5 \text{ N m}^{-2}$$

(*c*) If, as a further approximation, V is assumed to be equal to the volume of the solution, equation (2) becomes,

$$\Pi \approx RTc \quad \text{(van't Hoff equation)} \qquad (3)$$

where c is the concentration (n_B/V) of the solution. Therefore, for 1.00 mol kg^{-1} (i.e. 0.825×10^3 mol m^{-3}) aqueous sucrose solution at 20°C,

$$\Pi \approx 8.314 \times 293 \times 0.825 \times 10^3 \text{ N m}^{-2}$$
$$= 20.1 \times 10^5 \text{ N m}^{-2}$$

It can be seen that equation (2) (the Morse equation) gives a good approximation to the ideal osmotic pressure, whereas equation (3) (the van't Hoff equation) gives a poor approximation.

The experimental osmotic pressure of 27.2×10^5 N m^{-2} is greater than the ideal value of 24.1×10^5 N m^{-2}, which is consistent with a solute activity coefficient greater than unity.

EXAMPLE 5.10. EUTECTIC PHASE DIAGRAM

Silver (melting point 960°C) and copper (melting point 1083°C) form a eutectic at 779°C which contains 39·9 mole per cent copper. The system contains two solid solutions, α and β, of composition as follows:

Temperature/°C	Mole per cent copper	
	α	β
779	14·1	95·1
500	3·1	99·0
200	0·35	99·9

Assuming that the liquidus and solidus lines are straight, draw a temperature–composition phase diagram and determine for an alloy containing 20 mole per cent copper, (a) the temperature at which solid first appears, (b) the composition of the solid which first appears, (c) the proportion of solid in the alloy at 850°C, (d) the proportion of eutectic in the alloy at 779°C, (e) the composition of the alloy at 500°C.

The nature at different temperatures of an alloy containing 20 mole per cent copper is represented by the line XY *(Fig. 5.4)*.

(a) On cooling the liquid melt from high temperatures, solid first separates out at point a, which corresponds to 869°C.

(b) The composition of the solid which first appears is given by bc, i.e. 6·9 mole per cent Cu.

(c) On further cooling, the composition of the solid (solid solution α) follows line cd and that of the liquid follows line ae. Therefore, at 850°C,

$$\frac{\text{mole per cent of solid}}{\text{mole per cent of liquid}} = \frac{gh}{fg} = \frac{4\cdot2}{11\cdot5}$$

i.e. mole per cent of solid in alloy $= \dfrac{4\cdot2\times100}{4\cdot2+11\cdot5} = 26\cdot6$

(d) At 779°C (the eutectic temperature), there is no liquid present. The alloy contains solid solution α and a mixture of solid solution α and solid solution β of eutectic composition, such that,

$$\frac{\text{mole per cent of eutectic}}{\text{mole per cent of remainder of } \alpha} = \frac{di}{ie} = \frac{20\cdot0 - 14\cdot1}{39\cdot9 - 20\cdot0} = \frac{5\cdot9}{19\cdot9}$$

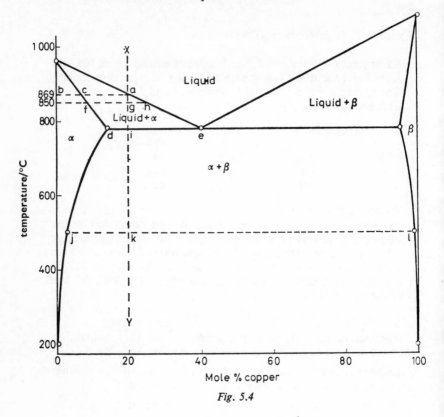

Fig. 5.4

i.e.   mole per cent of eutectic in alloy $= \dfrac{5 \cdot 9 \times 100}{5 \cdot 9 + 19 \cdot 9} = 22 \cdot 9$

(*e*) At 500°C,

$$\frac{\text{mole per cent of solid solution } \alpha}{\text{mole per cent of solid solution } \beta} = \frac{kl}{jk} = \frac{99 \cdot 9 - 20 \cdot 0}{20 \cdot 0 - 3 \cdot 1} = \frac{79 \cdot 9}{16 \cdot 9}$$

therefore, composition of alloy is $\dfrac{79 \cdot 9 \times 100}{79 \cdot 9 + 16 \cdot 9}$ mole per cent of solid solution α, i.e. 82·5 mole per cent α and 17·5 mole per cent β.

Additional Examples

1. From the following data obtain an approximate equation relating the saturated vapour pressure of water and the thermodynamic temperature:

Vapour pressure/N m^{-2}	2335	7375	19 190	47 300	101 300
Temperature/°C	20	40	60	80	100

2. The melting point of bismuth at normal pressure is 271·0°C. Under these conditions the densities of solid and liquid bismuth are 9·673 g cm^{-3} and 10·00 g cm^{-3}, respectively, and the melting point falls by 0·003 54 K when the pressure is increased by one atmosphere. Calculate the latent heat of fusion of bismuth.

[University of Salford, B.Sc. (lst year), 1967]

3. At 127°C HgI$_2$ undergoes a polymorphic change from the red to the yellow form. The heat of transition is 1250 J mol^{-1} and the change in molar volume on transition is 5·4 cm^3 mol^{-1}, the red form being the less dense. Calculate the rate of change of transition temperature with pressure at 127°C.

[University of Sheffield, B.Sc. (lst year), 1968]

4. The mean latent heat of vaporization of water between 90°C and 100°C is 2270 J g^{-1}. Calculate the vapour pressure of water at 90°C. (Liverpool Polytechnic, H.N.C., 1965)

5. The vapour pressure of diethyl ether is 0·247×10^5 N m^{-2} at 0°C and 1·228×10^5 N m^{-2} at 40°C. Calculate the latent heat of evaporation, assumed to be constant over this temperature range. At what temperature would ether boil at 1·013×10^5 N m^{-2} pressure? (G.R.I.C. Part 1, 1965)

6. The vapour pressure of chloroform varies with temperature as follows:

Temperature/°C	40	50	60	70	80
Vapour pressure/10^5 N m^{-2}	0·488	0·701	0·986	1·358	1·870

Determine the mean latent heat of evaporation of chloroform over this temperature range.

7. The vapour pressures of mercury at various temperatures are given in the following table:

T/°C	20	50	100	150	200	300	400
p/N m^{-2}	0·160	1·69	36·4	374	2300	32 900	210 000

Calculate (a) the molar enthalpy of evaporation of mercury at 25°C, and (b) the molar enthalpy and entropy of evaporation of mercury at its normal boiling point.

8. Estimate the volume of nitrogen (corrected to s.t.p.) dissolved in 1 dm³ of water in equilibrium with air at 20°C and standard atmospheric pressure. In the Henry law equation, $p_B = kx_B$, the constant, k, is 8.4×10^9 N m⁻² for nitrogen dissolved in water at 20°C.

9. The saturated vapour pressures at 50°C of CCl₄ and SiCl₄ are 0.424×10^5 N m⁻² and 0.800×10^5 N m⁻², respectively. Assuming ideal behaviour, calculate (a) the mole fraction of SiCl₄ in a mixture boiling at 50°C under an external pressure of 0.533×10^5 N m⁻², and (b) the mole fraction of SiCl₄ in the initial condensate formed on distilling this mixture.

10. Mixtures of *n*-hexane and *n*-heptane follow Raoult's law. At 330 K the vapour pressure of *n*-hexane is 0.68×10^5 N m⁻² and that of *n*-heptane is 0.24×10^5 N m⁻². Calculate the total vapour pressure of a solution containing 0.25 mole fraction of *n*-hexane and the mole fraction of *n*-hexane in the vapour in equilibrium with this solution. If some of this vapour is condensed to a liquid and a small amount is then allowed to re-evaporate so that liquid and vapour are again in equilibrium at 330 K, what will be the composition of the new vapour?

11. The table below gives values of the partial vapour pressures (in N m⁻²) of water and *n*-propanol in equilibrium with liquid mixtures of water and *n*-propanol as a function of the liquid composition at 25°C.

Mole fraction (n-propanol)	p (water)	p (n-propanol)
0·00	3170	0
0·02	3130	670
0·05	3030	1440
0·10	2930	1760
0·20	2910	1810
0·40	2890	1890
0·60	2650	2070
0·80	1790	2370
0·90	1080	2590
0·95	560	2770
1·00	0	2910

(a) From the vapour pressure–composition diagram show clearly where each component is behaving ideally (i.e. obeying Raoult's law).

(*b*) Calculate the composition of the vapour in equilibrium with a liquid mixture containing 0·1 mole fraction of *n*-propanol.

(*c*) What would the answer for (*b*) be if the liquid mixture behaved ideally over the whole composition range?

[University of Durham, B.Sc. (General), 1967]

12. The vapour pressure of the system, diethylaniline–water, is $1·013 \times 10^5$ N m^{-2} at 99·4°C. The vapour pressure of water at this temperature is $0·992 \times 10^5$ N m^{-2}. How many grammes of steam are necessary to distil over 100 g of diethylaniline?

[University of Birmingham, B.Sc. (1st year), 1966]

13. The distribution coefficient of a substance, X, between ether and water is 10, X having the greater solubility in ether. The maximum possible amount of X is ether-extracted from 100 cm^3 of an aqueous solution containing 10 g of X, (*a*) in one operation using 100 cm^3 of ether, (*b*) in five successive operations using 20 cm^3 of ether each time. Calculate the total weight of X which is ether-extracted from the aqueous solution in each case.

14. Some iodine is dissolved in a 0·1 mol dm^{-3} aqueous solution of potassium iodide which is then shaken with carbon tetrachloride until equilibrium is attained, the temperature being 15°C. The equilibrium concentrations of iodine are determined and found to be 0·050 mol dm^{-3} in the aqueous layer and 0·085 mol dm^{-3} in the carbon tetrachloride layer. The distribution coefficient of iodine between carbon tetrachloride and water is 85. Calculate the equilibrium constant for the reaction, $KI + I_2 = KI_3$, at 15°C.

15. A 0·2 mol kg^{-1} aqueous solution of a monobasic acid has a boiling point 0·150 K higher than that of pure water. If the latent heat of evaporation of water is 2250 J g^{-1}, calculate the dissociation constant of the acid consistent with these data.

[University of Nottingham, B.Sc. (Part 1), 1967]

16. Calculate the cryoscopic constant for water given that the enthalpy of fusion is 334 J g^{-1} at 273 K.

17. The freezing point of a solution of 0·55 g of nitrobenzene in 22 g of acetic acid is 0·78 K below that of pure acetic acid. If the freezing constant for acetic acid is 3·9 K mol^{-1} kg, calculate the relative molecular mass of nitrobenzene.

[University of Nottingham, B.Sc. (Inter.), 1966]

18. The freezing point of pure benzene is 5·40°C and its latent heat of fusion is 126·5 J g^{-1}. A solution containing 6·054 g of triphenylamine in 1 kg of benzene has a freezing point which is 0·126 3 K below that of the pure solvent. Calculate the relative molecular mass of the solute.

[Liverpool Polytechnic, B.Sc. (1st year), 1966]

19. The osmotic pressure (against pure water) of a solution containing 1 g of sucrose and y g of glucose in 1 kg of water at 25°C is 0.3×10^5 N m^{-2}. Calculate y, assuming ideal behaviour.

20. A solution of 1·02 g of a non-volatile solute in 100 g of chloroform has an osmotic pressure (against pure solvent) of 0.60×10^5 N m^{-2} at 20°C. What would be the vapour pressure of the solution at this temperature and what is the relative molecular mass of the solute? At 20°C, the density of liquid chloroform is 1·50 g cm^{-3} and its vapour pressure is 0.22×10^5 N m^{-2}.

[University of Birmingham, B.Sc. (lst year), 1967]

21. Antimony (m.p. 630°C) and lead (m.p. 326°C) form one eutectic mixture at 246°C which is 81 mole per cent lead, but do not form any solid solutions. Draw a temperature–composition diagram, assuming that the liquidus lines are linear, and label each region indicating which phases are in equilibrium under the conditions that the regions represent. For a mixture containing 50 mole per cent lead determine, (a) the temperature at which solid first crystallizes out, (b) the nature and proportion of solid in the mixture at 300°C, (c) the proportion of eutectic in the mixture at 200°C.

22. The temperatures at which solid first crystallizes out when liquid mixtures of benzophenone and diphenylamine are cooled are given in the following table:

Mole per cent diphenylamine	0	10	20	30	40		50	60	70	80	90	100
Temperature/°C						47	42	34	35	38·5	40	37	34	41	48	53

Draw a temperature–composition diagram and label each region indicating which phases are in equilibrium under the conditions that the regions represent. What conclusion can be drawn from the nature of the phase diagram?

CHAPTER 6

Reaction Kinetics

EXAMPLE 6.1. FIRST ORDER REACTION

The rate of a reaction was followed by measuring the absorbance of a solution at various times:

Time/min	0	18	57	130	240	337	398
Absorbance	1·39	1·26	1·03	0·706	0·398	0·251	0·180

Assuming that the Lambert–Beer law is obeyed, show that the reaction is first order and determine the rate constant.

The rate constant k of a first order reaction is given by

$$kt = 2 \cdot 303 \log_{10} \left(\frac{a}{a-x} \right)$$

where a is the initial concentration of reactant and $(a-x)$ is the concentration of reactant at time t.

Since the Lambert-Beer law is obeyed, absorbance is proportional to concentration, so that the first order equation becomes

$$kt = 2 \cdot 303 \log_{10} \frac{A_0}{A} = 2 \cdot 303 \log_{10} A_0 - 2 \cdot 303 \log_{10} A$$

where A_0 is the initial absorbance and A is the absorbance at time t.

A plot of $\log_{10} A$ against t should, therefore, be a straight line of slope $-\dfrac{k}{2\cdot303}$.

t/min	0	18	57	130	240	337	398
A	1·39	1·26	1·03	0·706	0·398	0·251	0·180
$\log_{10} A$	0·143	0·100	0·013	$\bar{1}$·849	$\bar{1}$·600	$\bar{1}$·400	$\bar{1}$·255

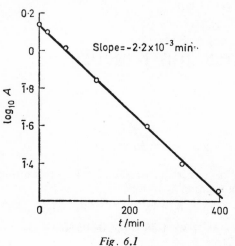

Fig. 6.1

Since the graph is a straight line, the reaction is first order, and

$$\text{slope of graph} = \frac{-k}{2\cdot303} = -2\cdot2\times10^{-3} \text{ min}^{-1}$$

i.e.
$$k = 5\cdot07\times10^{-3} \text{ min}^{-1}$$
$$= 8\cdot45\times10^{-5} \text{ s}^{-1}$$

EXAMPLE 6.2. PSEUDO-FIRST ORDER REACTION

A certain amount of methyl acetate was hydrolysed in the presence of an excess of 0·05 mol dm^{-3} hydrochloric acid at 25°C. When 25 cm^3 aliquots of reaction mixture were removed and titrated with NaOH solution, the volume V, of alkali required for neutralization after time, t, were as follows:

t/min	0	21	75	119	∞
V/cm^3	24·4	25·8	29·3	31·7	47·2

Show graphically or otherwise that the reaction is first order and calculate the time at which half the methyl acetate is hydrolysed.
[University of Nottingham B.Sc. (1st year), 1965]

The hydrolysis of methyl acetate in aqueous solution in the presence of hydrochloric acid, which acts as a catalyst, is a second order process, so that

$$\text{rate of reaction} = k_2[CH_3COOCH_3] \,[\text{acid}]$$

where k_2 is a second order rate constant, and the terms in square brackets are concentrations.

But since the acid is present in excess,

$$\text{rate of reaction} = k[CH_3COOCH_3]$$

where k is the pseudo-first order rate constant, and is equal to k_2 [acid].

The titre, V, of alkali is a measure of the concentration of acid catalyst and the concentration of acetic acid formed as a product.

If V_∞ is the titre when the reaction is completed, $V_\infty - V_0$ will be proportional to the initial concentration of ester a. Similarly, if V_t is the titre at time t, $V_\infty - V_t$ is proportional to the ester concentration at time t.

The first order rate constant is, therefore, given by

$$kt = 2 \cdot 303 \log_{10}\left(\frac{V_\infty - V_0}{V_\infty - V_t}\right)$$

i.e. $kt = 2 \cdot 303 \log_{10}(V_\infty - V_0) - 2 \cdot 303 \log_{10}(V_\infty - V_t)$

If the reaction is first order, a plot of $\log_{10}[V_\infty - V_t]$ against t should be a straight line of slope $-\dfrac{k}{2 \cdot 303}$ *(Fig. 6.2)*

t/min	0	21	75	119	∞
V_t/cm³	24·4	25·8	29·3	31·7	47·2
$(V_\infty - V_t)$/cm³	22·8	21·4	17·9	15·5	−
$\log_{10}[(V_\infty - V_t)$/cm³]	1·358	1·330	1·253	1·190	−

Since the graph is a straight line, the reaction is first order, and the slope is

$$-k/2 \cdot 303 = -1 \cdot 46 \times 10^{-3} \text{ min}^{-1}$$

i.e. $k = 3 \cdot 36 \times 10^{-3} \text{ min}^{-1}$

$= 5 \cdot 60 \times 10^{-5} \text{ s}^{-1}$

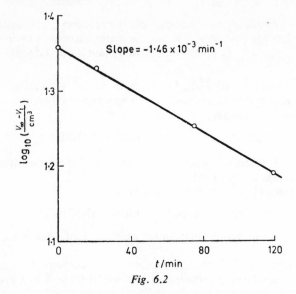

Fig. 6.2

At the half-life, $t_{0.5}$, the reaction will have proceeded to half completion, and the titre will be $0.5(V_\infty - V_0)$.

Therefore,

$$kt_{0.5} = 2.303 \log_{10}\left[\frac{V_\infty - V_0}{0.5(V_\infty - V_0)}\right]$$

$$= 2.303 \log_{10} 2 = 0.693$$

from which,

$$t_{0.5} = \frac{0.693}{5.60 \times 10^{-5} \text{ s}^{-1}} = 1.24 \times 10^4 \text{ s}$$

Note. The half-life of any first order reaction is independent of the initial concentration, and is given by $t_{0.5} = 0.693/k$.

EXAMPLE 6.3. SECOND ORDER REACTION

The kinetics of the reaction between methyl p-toluenesulphonate and sodium iodide in acetone solution at $26.5°C$ were followed by a titration method. Equal initial concentrations of each reactant were used, and the concentration of each reactant at the following times is given:

Concentration/mol dm^{-3} × 10^{-2}	5·00	4·85	4·72	4·48	4·26
Time/h	0	0·5	1	2	3

Concentration/mol dm^{-3} × 10^{-2}	4·03	3·86	3·70	3·55	3·40
Time/h	4	5	6	7	8

Determine the order and calculate the rate constant for the reaction.

Since the concentration of both reactants varies with time in the same way, a second order reaction is indicated.

For a second order reaction, where the initial concentration, a, of each reactant is the same, the rate constant, k, is given by

$$kt = \frac{1}{(a-x)} - \frac{1}{a}$$

where $(a-x)$ is the concentration at time t.

A plot of $\dfrac{1}{(a-x)}$ against t should, therefore, be a straight line of slope, k, if the reaction is second order *(Fig. 6.3)*.

Time/h	$\dfrac{(a-x)}{\text{mol dm}^{-3}}$	$\dfrac{\text{mol dm}^{-3}}{(a-x)}$
0	0·050 0	20·0
0·5	0·048 5	20·6
1	0·047 2	21·2
2	0·044 8	22·3
3	0·042 6	23·5
4	0·040 3	24·8
5	0·038 6	25·9
6	0·037 0	27·0
7	0·035 5	28·2
8	0·034 0	29·4

Since the graph *(Fig. 6.3)* is a straight line, the reaction is second order, and

$$\text{slope} = k = 1·177 \text{ dm}^3 \text{ mol}^{-1} \text{ h}^{-1}$$
$$= 3·27 \times 10^{-4} \text{ dm}^3 \text{ mol}^{-1} \text{ s}^{-1}$$

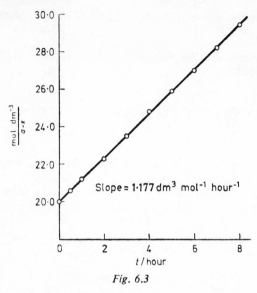

Fig. 6.3

EXAMPLE 6.4. THIRD ORDER REACTION

The reaction between nitric oxide and hydrogen

$$2\,NO + 2\,H_2 \rightarrow N_2 + 2\,H_2O$$

has been investigated by measuring the initial rate of decrease of pressure in known mixtures of gases. The following results were obtained at 700°C.

| Initial pressure/atm | | Initial rate of decrease |
NO	H_2	of pressure/atm min^{-1}
0·5	0·2	0·004 8
0·5	0·1	0·002 4
0·25	0·2	0·001 2

Deduce the order of the reaction with respect to each reactant. Calculate the rate constant for the overall reaction at 700°C.
[Liverpool Polytechnic B.Sc. (2nd year), 1968]

The initial rate of decrease of pressure is given by

$$-\frac{\mathrm{d}p}{\mathrm{d}t} = k p_{NO}^{x} p_{H_2}^{y}$$

where p_{NO} and p_{H_2} are the initial pressures of NO and H_2 respectively, x and y are the order with respect to NO and H_2, and k is the rate constant expressed in pressure units.

Let the initial rate of decrease of pressure be represented by v. Taking logarithms,

$$\log_{10} v = \text{constant} + x \log_{10} p_{NO} + y \log_{10} p_{H_2}$$

For the first two sets of data, p_{NO} is constant so that

$$\log_{10} v = \text{constant} + y \log_{10} p_{H_2}$$

or

$$\log_{10} \frac{v_1}{v_2} = y \log_{10} \frac{(p_{H_2})_1}{(p_{H_2})_2}$$

where v_1 is the rate at initial pressure $(p_{H_2})_1$ and v_2 is the rate at initial pressure $(p_{H_2})_2$.

Substituting the appropriate numerical values,

$$\log_{10} \left(\frac{0 \cdot 004\ 8}{0 \cdot 002\ 4} \right) = y \log_{10} \left(\frac{0 \cdot 2}{0 \cdot 1} \right)$$

giving

$$y = 1$$

Similarly for the first and last set of data p_{H_2} is constant, so that

$$\log_{10} \frac{v_1}{v_3} = y \log_{10} \frac{(p_{NO})_1}{(p_{NO})_3}$$

where v_1 and v_3 are the rates when the initial pressure of NO is $(p_{NO})_1$ and $(p_{NO})_3$ respectively.

Substituting the appropriate numerical values,

$$\log_{10} \left(\frac{0 \cdot 004\ 8}{0 \cdot 001\ 2} \right) = x \log_{10} \left(\frac{0 \cdot 5}{0 \cdot 25} \right)$$

giving

$$x = 2$$

The reaction is, therefore, overall third order, and the rate equation is

$$-dp/dt = k(p_{NO})^2 p_{H_2}$$

From the first set of data, the rate constant, k, is given by

$$k = \frac{dp/dt}{(p_{NO})^2 p_{H_2}} = \frac{0 \cdot 004\ 8}{(0 \cdot 5)^2 \times 0 \cdot 2} \text{ atm}^{-2} \text{ min}^{-1}$$
$$= 0 \cdot 096 \text{ atm}^{-2} \text{ min}^{-1}$$
$$= \frac{0 \cdot 096}{60 \times (1 \cdot 013 \times 10^5)^2} (\text{N m}^{-2})^{-2} \text{ s}^{-1}$$
$$= 1 \cdot 56 \times 10^{-13} \text{ m}^4 \text{ N}^{-2} \text{ s}^{-1}$$

Note. The rate constant can be expressed in concentration units if it is assumed that the gases behave ideally, i.e. $c = p/RT$, where c is the concentration, p is the pressure, R is the gas constant and T is the thermodynamic temperature. The rate constant on the basis of concentration units is, therefore, given by

$$k = \frac{\mathrm{d}(p/RT)}{\mathrm{d}T} \bigg/ \left(\frac{p_{NO}}{RT}\right)^2 \left(\frac{p_{H_2}}{RT}\right)$$

$$= 1 \cdot 56 \times 10^{-13} \times (8 \cdot 314)^2 \times (973)^2 \ \mathrm{m}^6 \ \mathrm{mol}^{-2} \ \mathrm{s}^{-1}$$

$$= 1 \cdot 02 \times 10^{-5} \ \mathrm{m}^6 \ \mathrm{mol}^{-2} \ \mathrm{s}^{-1}$$

EXAMPLE 6.5. RATE EQUATION FROM HALF-LIFE MEASUREMENTS

The acid catalysed hydrolysis of an organic compound A at 30°C has a time for half-change of 100 min when carried out in a buffer solution at pH 5, and 10 min when carried out at pH 4. Both times of half-change are independent of the initial concentration of A. If the rate constant, k, is given by

$$\frac{-\mathrm{d}[A]}{\mathrm{d}t} = k[A]^a \, [H^+]^b$$

what are the values of a and b?

[University of Bristol B.Sc. (Ord.), 1966]

The rate equation for the above reaction is

$$\frac{-\mathrm{d}\,[A]}{\mathrm{d}t} = k[A]^a[H^+]^b$$

During any experiment, $[H^+]$ is constant, therefore

$$\frac{-\mathrm{d}\,[A]}{\mathrm{d}t} = k'[A]^a$$

where $k' = k[H^+]^b$.

Since the half-life is independent of the initial concentration of A, the rate has a first order dependence on $[A]$, i.e. $a = 1$. Consequently k' is a first order rate constant and is given by $k' = 0 \cdot 693/t_{0 \cdot 5}$.

Therefore,

$$\frac{(t_{0 \cdot 5})_1}{(t_{0 \cdot 5})_2} = \frac{k_2'}{k_1'} = \frac{[H^+]_2^b}{[H^+]_1^b}$$

Substituting the appropriate numerical values for the dependence of half-life on hydrogen ion concentration,

$$\frac{100}{10} = \left(\frac{10^{-4}}{10^{-5}}\right)^{b}$$

i.e. $\qquad\qquad b = 1$

Therefore the rate of hydrolysis has a first order dependence on $[H^{+}]$, so that

$$-\frac{d[A]}{dt} = k[A][H^{+}]$$

EXAMPLE 6.6. FRACTIONAL DECOMPOSITION

The following table gives the extent of decomposition of vinyl ethyl ether at 389°C at various times for an initial pressure of 51 torr.

Per cent decomposition	20	30	40	50
Time/s	264	424	609	820

Calculate the rate constant of the reaction.

For a first order reaction, the rate constant, k, is given by

$$kt = 2 \cdot 303 \log_{10}\left(\frac{a}{a-x}\right)$$

where a is the initial concentration and x is the decrease in concentration at time t.

Let t be the time taken for the concentration to decrease to a fraction x/a of the initial concentration.
Then

$$kt = 2 \cdot 303 \log_{10}\left(\frac{1}{1-x/a}\right)$$

A plot of $\log_{10}\left(\frac{1}{1-x/a}\right)$ against t should, therefore, give a straight line of slope $k/2 \cdot 303$, if the reaction is first order.

t/s	Per cent decomposition	Fraction decomposition (x/a)	$\dfrac{1}{1-x/a}$	$\log_{10}\left(\dfrac{1}{1-x/a}\right)$
264	20	0·2	1·250	0·097
424	30	0·3	1·429	0·155
609	40	0·4	1·667	0·222
820	50	0·5	2·000	0·301

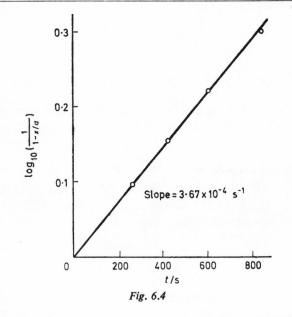

Fig. 6.4

Since the graph is a straight line, the reaction is first order, and

$$\text{slope} = \frac{k}{2\cdot303} = 3\cdot67\times10^{-4}\ \text{s}^{-1}$$

i.e. $k = 8\cdot45\times10^{-4}\ \text{s}^{-1}$

EXAMPLE 6.7. GAS PHASE DECOMPOSITION

At 155°C the gas-phase decomposition of di-*t*-butyl peroxide is a first order reaction given by

$$(CH_3)_3COOC(CH_3)_3 \rightarrow 2(CH_3)_2CO + C_2H_6$$

The following table gives the results obtained by Ralley *et al.* [*J. Am. Chem. Soc.*, 70 (1948) 88] at time, t, for the total pressure, p, measured on a fixed volume:

t/min	0	3	6	9	12	15	18	21
p/Torr	169·3	189·2	207·1	224·4	240·2	255·0	269·7	282·6

Calculate the rate constant for the reaction.

At any time, the total pressure, p, is given by

$$p = p_{DTBP} + p_A + p_E$$

where p_{DTBP}, p_A and p_E are the pressures of peroxide, acetone and ethane, respectively.

If y is the decrease in the pressure of peroxide in time t from an initial pressure of p_0, then

$$p_{DTBP} = p_0 - y$$
$$p_A = 2y$$

and

$$p_E = y$$

Therefore

$$p = p_0 + 2y$$

i.e.

$$y = \frac{p - p_0}{2}$$

and

$$p_{DTBP} = \frac{3p_0 - p}{2}$$

The first order rate equation is given by

$$kt = 2 \cdot 303 \log_{10} \left(\frac{a}{a-x} \right)$$

The initial concentration, a, is proportional to p_0, and the concentration $(a-x)$, at time, t, is proportional to $\dfrac{3p_0 - p}{2}$

The rate constant, k, is, therefore, given by

$$kt = 2 \cdot 303 \log_{10} \left(\frac{2p_0}{3p_0 - p} \right)$$

$$= 2 \cdot 303 \log_{10} 2p_0 - 2 \cdot 303 \log_{10} (3p_0 - p)$$

A plot of $\log_{10} (3p_0 - p)$ against t should, therefore, be a straight line of slope $-\dfrac{k}{2 \cdot 303}$.

Time/min	p/Torr	$(3p_0 - p)$/Torr	$\log_{10}[(3p_0 - p)/\text{Torr}]$
0	169·3	338·6	2·529 7
3	189·2	318·7	2·503 4
6	207·1	300·8	2·478 2
9	224·4	283·5	2·452 6
12	240·2	267·7	2·427 7
15	255·0	252·9	2·402 9
18	269·7	238·2	2·377 0
21	282·6	225·3	2·352 8

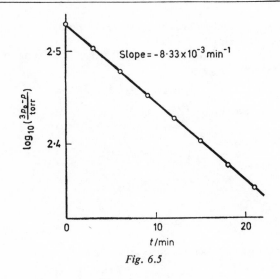

Fig. 6.5

From the graph *(Fig. 6.5)*,

$$\text{slope} = -\frac{k}{2 \cdot 303} = -8 \cdot 33 \times 10^{-3} \text{ min}^{-1}$$

i.e.
$$k = 1 \cdot 92 \times 10^{-2} \text{ min}^{-1}$$
$$= 3 \cdot 20 \times 10^{-4} \text{ s}^{-1}$$

EXAMPLE 6.8. COLLISION THEORY

The bimolecular decomposition of hydrogen iodide is given by the equation

$$2\,HI \rightarrow H_2 + I_2$$

Assuming a collision diameter of 0·35 nm for HI and an activation energy of 184 kJ mol^{-1} for the reaction, calculate (a) the collision rate, (b) the reaction velocity and (c) the rate constant, for the above reaction at 700 K and one atmosphere pressure.

(a) From kinetic theory, the collision rate Z (i.e. the number of collisions per unit volume of gas per unit time) is given by

$$Z = 2n^2\sigma^2 \sqrt{\left(\frac{\pi RT}{M}\right)}$$

where n is the number of molecules per unit volume of gas, σ is the collision diameter, R is the gas constant, T is the thermodynamic temperature and M is the molar mass.

Assuming that HI behaves as an ideal gas,

$$n = \frac{pN_A}{RT}$$

where p is the pressure and N_A is the Avogadro constant. Substituting,

$$n = \frac{(1·013\times 10^5)\times 1\times(6·023\times 10^{23})}{8·314\times 700} \text{ molecule m}^{-3}$$

$$= 1·05\times 10^{25} \text{ molecule m}^{-3}$$

Therefore,

$$Z = 2\times(1·05\times 10^{25})^2\times(0·35\times 10^{-9})^2 \sqrt{\left(\frac{\pi\times 8·314\times 700}{127·9\times 10^{-3}}\right)} \text{ m}^{-3}\text{ s}^{-1}$$

$$= 1·02\times 10^{34} \text{ m}^{-3}\text{ s}^{-1}$$

(b) From the simple collision theory of reaction rates, the reaction velocity, v, is given by

$$v = 2Z \exp(-\Delta E^+/RT)$$

where ΔE^+ is the activation energy, and the factor of 2 is introduced because 2 molecules of HI are involved in each collision.

Substituting,

$$v = 2\times(1·02\times 10^{34})\exp(-184\,000/8·314\times 700) \text{ molecule m}^{-3}\text{ s}^{-1}$$

Therefore,

$$\log_{10}(v/\text{molecule m}^{-3}\text{ s}^{-1}) = 34·310 - \frac{184\,000}{2·303\times 8·314\times 700}$$

$$= 20·574$$

giving, $\qquad v = 3·75\times 10^{20} \text{ molecule m}^{-3}\text{ s}^{-1}$

Dividing by Avogadro's constant,

$$v = \frac{3.75 \times 10^{20}}{6.023 \times 10^{23}} \text{ mol m}^{-3} \text{ s}^{-1} = 6.23 \times 10^{-4} \text{ mol m}^{-3} \text{ s}^{-1}$$

$$= 6.23 \times 10^{-7} \text{ mol dm}^{-3} \text{ s}^{-1}$$

(c) Since the rate constant, k, is given by $v = kn^2$,

$$k = \frac{v}{n^2} = \frac{3.75 \times 10^{20}}{(1.05 \times 10^{25})^2} \text{ m}^3 \text{ molecule}^{-1} \text{ s}^{-1}$$

$$= 3.40 \times 10^{-30} \text{ m}^3 \text{ molecule}^{-1} \text{ s}^{-1}$$

Multiplying by Avogadro's constant,

$$k = 2.05 \times 10^{-6} \text{ m}^3 \text{ mol}^{-1} \text{ s}^{-1} = 2.05 \times 10^{-3} \text{ dm}^3 \text{ mol}^{-1} \text{ s}^{-1}$$

Note. The rate constant can be calculated directly from the data given, since,

$$k = \frac{v}{n^2} = \frac{2Z \exp\left(-\Delta E^{\ast}/RT\right)}{n^2}$$

from which,

$$k = 4\sigma^2 \sqrt{\left(\frac{\pi RT}{M}\right)} \exp\left(-\Delta E^{\ast}/RT\right)$$

EXAMPLE 6.9. ENERGY OF ACTIVATION

The rate constants for the first order decomposition of acetone dicarboxylic acid in aqueous solution are found to have the following values at the temperature stated:

Temperature/°C	0	10	20	30	40	50	60
Rate constant $\times 10^5/\text{s}^{-1}$		2.46	10.8	47.5	163	576	1850	5480

Calculate the energy of activation for the reaction.

(G.R.I.C. Part 1, 1963)

The Arrhenius equation expresses the rate constant, k, in terms of the frequency factor, A, and the activation energy, $\Delta E^{\ast}$, by

$$k = A \exp\left(-\Delta E^{\ast}/RT\right)$$

Taking logarithms

$$\ln k = \ln A - \frac{\Delta E^{\ddagger}}{RT}$$

i.e.

$$\log_{10} k = \log_{10} A - \frac{\Delta E^{\ddagger}}{2 \cdot 303 RT}$$

A plot of $\log_{10} k$ against $\frac{1}{T}$ should, therefore, give a straight line of

slope $-\dfrac{\Delta E^{\ddagger}}{2 \cdot 303 R}$

$k \times 10^5/\text{s}^{-1}$	2·46	10·8	47·5	163	576	1850	5480
$\log_{10}(k/\text{s}^{-1})$	$\bar{5}$·391	$\bar{4}$·033	$\bar{4}$·677	$\bar{3}$·212	$\bar{3}$·760	$\bar{2}$·267	$\bar{2}$·739
T/K	273	283	293	303	313	323	333
$\dfrac{10^3 \text{ K}}{T}$	3·663	3·534	3·413	3·300	3·195	3·096	3·003

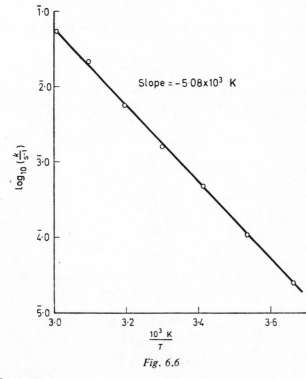

Fig. 6.6

From the slope of the graph *(Fig. 6.6)*

$$-\frac{\Delta E^{\ddagger}}{2 \cdot 303 R} = -5 \cdot 08 \times 10^3 \text{ K}$$

i.e.
$$\Delta E^{\ddagger} = 2 \cdot 303 \times 8 \cdot 314 \times 5 \cdot 08 \times 10^3 \text{ J mol}^{-1}$$
$$= 97 \cdot 3 \text{ kJ mol}^{-1}$$

EXAMPLE 6.10. HALF–LIFE AND ENERGY OF ACTIVATION

For a particular first order reaction at 27°C, the concentration of reactant is reduced to one half of its initial value after 5000 s. At 37°C, the concentration is halved after 1000 s. Calculate (a) the rate constant of the reaction at 27°C, (b) the time required for the concentration to be reduced to one quarter of its initial value at 37°C, (c) the activation energy of the reaction.

[University of Birmingham. B.Sc. (1st year), 1966]

(a) The rate constant, k, of a first order reaction is given by

$$kt = 2 \cdot 303 \log_{10}\left(\frac{a}{a-x}\right)$$

where a is the initial concentration and $a-x$ is the concentration at time t. The half life, $t_{0.5}$, is the time taken to reduce the initial concentration by half, i.e. $x = a/2$, so that

$$kt_{0.5} = 2 \cdot 303 \log_{10} 2 = 0 \cdot 693$$

Therefore,
$$k(27°C) = \frac{0 \cdot 693}{5000 \text{ s}}$$
$$= 1 \cdot 39 \times 10^{-4} \text{ s}^{-1}$$

(b) Similarly,
$$k(37°C) = \frac{0 \cdot 693}{1000 \text{ s}}$$
$$= 6 \cdot 93 \times 10^{-4} \text{ s}^{-1}$$

The time, $t_{0.25}$ required to reduce the initial concentration, a, to one quarter of its initial value at 37°C is given by

$$t_{0.25} = \frac{2 \cdot 303 \log_{10}\left(\frac{a}{0 \cdot 25a}\right)}{6 \cdot 93 \times 10^{-4} \text{ s}^{-1}}$$
$$= 2000 \text{ s (i.e. } 2 \times t_{0.5})$$

(c) The rate constant, k, is given by

$$k = A \exp\left(-\Delta E^{\ddagger}/RT\right)$$

where A is a constant known as the frequency factor, $\Delta E^{\ddagger}$ is the activation energy, and T is the thermodynamic temperature. Taking logarithms,

$$\ln k = \ln A - \frac{\Delta E^{\ddagger}}{RT}$$

Assuming that $\Delta E^{\ddagger}$ is constant over the temperature range,

$$\ln\left(\frac{k_2}{k_1}\right) = -\frac{\Delta E^{\ddagger}}{R}\left(\frac{1}{T_2} - \frac{1}{T_1}\right)$$

where k_1 and k_2 are the rate constants at temperatures T_1 and T_2 respectively, i.e.

$$\log_{10}\left(\frac{k_2}{k_1}\right) = +\frac{\Delta E^{\ddagger}}{2\cdot303R}\left(\frac{T_2 - T_1}{T_1 T_2}\right)$$

Substituting the appropriate numerical values,

$$\log_{10}\left(\frac{6\cdot93\times10^{-4}}{1\cdot39\times10^{-4}}\right) = \frac{\Delta E^{\ddagger}/\text{J mol}^{-1}}{2\cdot303\times8\cdot314}\left(\frac{310-300}{300\times310}\right)$$

i.e. $\quad \Delta E^{\ddagger} = \dfrac{0\cdot697\,7\times2\cdot303\times8\cdot314\times300\times310}{10}\,\text{J mol}^{-1}$

$$= 124\ \text{kJ mol}^{-1}$$

Additional Examples

1. The following data were obtained for the hydrolysis of a sugar in aqueous solution at 27°C:

Time/min	0	60	130	180
Sugar concentration/ mol dm^{-3}	1·000	0·807	0·630	0·531

Show that the reaction is first order and calculate the rate constant for the hydrolysis.

2. The following data refer to the decomposition of benzene diazonium chloride

$$C_6H_5N_2Cl \rightarrow C_6H_5Cl + N_2$$

at a starting concentration of 10 g dm^{-3} in solution at 50°C:

Time/min	6	9	12	14	18	22	24	26	30	∞
N$_2$ *evolved*/cm^3	19·3	26·0	32·6	36·0	41·3	45·0	46·5	48·4	50·4	58·3

Find the order, the rate constant and the half-life of the reaction.

(G.R.I.C. Part 1, 1967)

3. The kinetics of the alcoholysis of cinnamal chloride was studied spectroscopically and gave the following results for absorbance as a function of time:

Absorbance	0·378	0·339	0·295	0·246	0·204	0·178
Time/min	0	20	40	70	100	120
Absorbance	0·148	0·107	0·006			
Time/min	150	200	∞			

Calculate the rate constant for the reaction.

4. In the spectroscopic study of the first order isomerization of *cis*-biethylene-diamine-dichloro-cobalt-III-chloride in methanol, the rate of disappearance of the absorption peak at 540 nm was followed as a function of time:

Absorbance	0·119	0·115	0·108	0·102	0·096		
Time/min	0	10	20	33	47		
Absorbance	0·089	0·081	0·075	0·071	0·066	0·058	0·005
Time/min	62	80	93	107	121	140	∞

Show that the reaction is first order and calculate the half-life of the reaction.

5. The acid catalysed hydrolysis of acetal at 35°C

$$CH_3CH(OC_2H_5)_2 + H_2O \rightarrow CH_3CHO + 2 C_2H_5OH$$

in excess water is a first order process. The change in volume during the reaction was measured with a dilatometer with the following results:

Dilatometer reading/cm	0	97	113	147	159	171	179	184	187
Time/min	0	3	5	8	10	12	15	20	∞

Calculate the rate constant and the half-life of the reaction.

6. Two substances A and B undergo a bimolecular reaction step. The following table gives the concentrations of A at various times for an experiment carried out at a constant temperature of 17°C:

Concentration of A×

10^4/mol dm^{-3}	10·00	7·94	6·31	5·01	3·98
Time/min	0	10	20	30	40

The initial concentration of B is 2·5 mol dm^{-3}. Calculate the value of the second order rate constant.

[University of Manchester, B.Sc. (1st year), 1967]

7. In studying the rate of a decomposition reaction it was found that the concentration of the reactant decreased from its initial value of 0·8 mol dm^{-3} in the following way:

Reactant concentration

/mol dm^{-3}	0·6	0·4	0·2	0·1
Time/min	4·17	12·5	37·5	87·5

What is the order of the reaction and what is the rate constant?

[University of Sheffield. B.Sc. (1st year), 1967]

8. The saponification of ethyl acetate in sodium hydroxide solution at 30°C

$$CH_3COOC_2H_5 + NaOH \rightarrow CH_3COONa + C_2H_5OH$$

was studied by Smith and Lorenson [*J. Am. Chem. Soc.*, 61 (1939) 117]. The initial concentrations of ester and alkali were both 0·05 mol dm^{-3}, and the decrease in ester concentration, x, was measured at the following times:

x/mol dm^{-3}

0·005 91	0·011 42	0·016 30	0·022 07	0·027 17	0·031 47	0·036 44

Time/min

4	9	15	24	37	53	83

Calculate the rate constant of the reaction.

9. When the concentration of A in the simple reaction

$$A \rightarrow B$$

was changed from 0·502 mol dm^{-3} to 1·007 mol dm^{-3}, the half-life dropped from 51 s to 26 s at 26°C. What is the order of the reaction and the value of the rate constant?

[University of Sheffield, B.Sc. (1st year), 1967]

10. Ethyl acetate and sodium hydroxide in solution in an alcohol–water mixture at 30°C exhibits a hydrolysis reaction. In an experiment in which $5·0 \times 10^{-2}$ mol dm^{-3} of each reactant were present at $t = 0$, the time of half-change was 1800 s and the time for three-quarters change was 5400 s. Deduce the order of the reaction

and calculate the rate constant. At what time was 10 per cent of the reaction completed?

[University of Bristol, B.Sc. (Part 1), 1966]

11. In the acid-catalysed hydrolysis of ethyl propionate, 14·1 per cent of the ester was found to have hydrolysed after 30 min and 70·3 per cent after 240 min. Show that the reaction is first order, find the velocity constant for the reaction, and determine the time of half-change.

[University of Bristol, B.Sc. (Subsid.), 1967]

12. At 585°C, the decomposition of nitrous oxide into nitrogen and oxygen obeys a first order law. The pressure exerted during the decomposition of nitrous oxide at constant volume was measured at various times, with the following results:

Pressure/torr	374·3	415·4	435·0	455·9	468·5
Time/h	0	26·5	42·0	62·6	75·0

Calculate the rate constant at 585°C.

13. The gas phase decomposition of ethane

$$C_2H_6 \rightarrow C_2H_4 + H_2$$

at 856 K was investigated by following the change in total pressure with time at constant volume, with the following results:

Total pressure/torr	384	390	394	396	400	405	408
Time/s	0	29	50	64	84	114	134

Determine the order of the reaction and the rate constant at this temperature.

14. The decomposition of *n*-propyl chloride occurs according to the stoichiometry

$$CH_3CH_2CH_2Cl \rightarrow CH_3CH = CH_2 + HCl$$

and has been studied by measurement of the pressure increase in a constant volume system. Starting with an initial pressure of 112 torr of pure *n*-propyl chloride at 713 K, the following pressures, p, were observed at the times, t, specified:

Time (t)/min	15	30	45	60	75
Pressure (p)/torr	136	155	170	181	191

Confirm that the decomposition obeys first order kinetics and calculate the specific rate constant.

[University of Liverpool, B.Sc. (Part 2), 1967]

15. In the thermal decomposition of a gaseous substance A into gaseous products, the following changes in total pressure are observed:

Time/min	0	5	10	15	30	∞
Total pressure/torr	200	267	300	320	350	400

Find the order of the reaction.

16. The rate of decomposition of nitrous oxide (N_2O) at the surface of an electrically-heated gold wire has been followed manometrically with the following results at 990°C:

Time/min	0	30	52	100
Pressure/torr	200	232	250	272

(*a*) Assuming the products are exclusively nitrogen and oxygen, calculate the final pressure and the half-life.

(*b*) At the same temperature but at an initial pressure of 400 torr, the half-life was found to be 52 min. What is the order of the reaction?

[University of Birmingham, B.Sc. (1st year), 1967]

17. The rate constant, k, for the alkaline hydrolysis of ethyl iodide was measured at various temperatures as shown:

Temperature/°C	15	30	60	90
$10^3 k$/dm³ mol^{-1} s^{-1}	0·0507	0·335	8·13	119

Use a graphical method to calculate the activation energy of the reaction, and determine the frequency factor.

18. The rate constant for the first order decomposition of 2-chloropropane into propylene and hydrogen chloride was found to vary with temperature as follows:

Rate constant $\times 10^3$/s^{-1}

0·162	0·238	0·311	0·475	0·706	0·901	1·225	1·593

Temperature/K

640·6	646·7	651·2	657·5	665·1	669·0	674·9	679·7

Calculate the energy of activation and the frequency factor.

19. The unimolecular isomerization of hepta-1,2,6-triene to 2-vinyl penta-1,4-diene

$$CH_2{=}C{=}CH{-}CH_2{-}CH_2{-}CH{=}CH_2$$
$$\rightarrow CH_2{=}CH{-}\underset{\underset{\displaystyle CH_2}{\|}}{C}{-}CH_2{-}CH{=}CH_2$$

has recently been shown to obey a first order law between 170°C and 220°C. The velocity constant varies with temperature as follows:

Temperature/°C   172·2   187·7   202·6   218·1

Rate constant/s^{-1}

$0·997 \times 10^{-4}$ $3·01 \times 10^{-4}$ $7·80 \times 10^{-4}$ $20·4 \times 10^{-4}$

Calculate the activation energy and the pre-exponential factor.
[University of Bristol, B.Sc. (Ord.), 1967]

20. The neutralization reaction of nitroethane in aqueous alkaline solution proceeds according to the rate equation:

$$-d[OH^-]/dt = -d[C_2H_5NO_2]/dt = k[C_2H_5NO_2][OH^-]$$

Experiments at 0°C with initial concentrations $[C_2H_5NO_2] = 0·01$ mol dm^{-3} and $[NaOH] = 0·01$ mol dm^{-3} give a value of 150 s for the reaction half-life. Calculate the corresponding rate constant at 0°C.

Experiments at 25°C gave a value of 5·90 dm^3 mol^{-1} s^{-1} for the reaction rate constant; calculate the activation energy of the reaction.

(G.R.I.C. Part 1, 1968)

21. The initial stage of the reaction between gaseous ammonia and nitrogen dioxide follows second order kinetics. Given that the rate constant at 600 K is $3·85 \times 10^2$ cm^3 mol^{-1} s^{-1} and at 716 K is $1·60 \times 10^4$ cm^3 mol^{-1} s^{-1}, calculate the activation energy and the Arrhenius pre-exponential factor.

22. In the homogeneous decomposition of nitrous oxide, it is found that the time for 50 per cent decomposition at constant temperature is inversely proportional to the initial pressure. Deduce the order of the reaction.

On varying the temperature, the following half-lives are obtained:

Temperature/°C	694	757
Initial pressure/torr	294	360
Half-life/s	1520	212

Calculate (*a*) the rate constants at 694°C and 757°C, (*b*) the activation energy, (*c*) the Arrhenius '*A*' factor.
State clearly the units in which your answers are expressed.
[University of Manchester, B.Sc. (1st year), 1967]

23. Two second order reactions A and B have identical frequency factors. The activation energy of A exceeds that of B by

10·46 kJ mol^{-1}. At 100°C, the reaction A is 30 per cent complete after 60 min when the reactant is initially present at a concentration of 0·1 mol dm^{-3}. How long will it take reaction B to reach 70 per cent completion at the same temperature for an initial concentration of 0·05 mol dm^{-3}?

[University of Birmingham, B.Sc. (lst year), 1967]

24. The decomposition of compound A in solution is a first orde r process with an activation energy of 52·3 kJ mol^{-1}. A 10 per cent solution of A is 10 per cent decomposed in 10 min at 10°C. How much decomposition would be obseřved with a 20 per cent solution after 20 min at 20°C?

(G.R.I.C. Part 1, 1969)

25. The hydrolysis of t-butyl bromide in aqueous acetone is a first order reaction which may be represented by the equation

$$t\text{-}C_4H_9Br + H_2O \rightarrow t\text{-}C_4H_9OH + HBr$$

From the following experimental results, calculate the rate constant of the reaction at both temperatures, and hence calculate the energy of activation for the reaction.

	25°C		50°C
Time/min	[t-C$_4$H$_9$Br]/mol dm^{-3}	*Time*/min	[t-C$_4$H$_9$Br]/mol dm^{-3}
0	0·103 9	0	0·105 6
195	0·089 6	18	0·085 6
380	0·077 6	40	0·064 5
600	0·063 9	72	0·043 2

[University of Salford, B.Sc. (Part 1), 1967]

26. In a certain reaction, the rate constant at 35°C is double its value at 25°C. Calculate the activation energy.

27. The Arrhenius equations for the rate of decomposition of methyl nitrite and ethyl nitrite are

$$k_1/s^{-1} = 10^{13} \exp\left(\frac{-152\,300 \text{ J mol}^{-1}}{RT}\right)$$

and

$$k_2/s^{-1} = 10^{14} \exp\left(\frac{-157\,700 \text{ J mol}^{-1}}{RT}\right)$$

respectively. Find the temperature at which the rate constants are equal.

CHAPTER 7

Electrochemistry

EXAMPLE 7.1. MOLAR CONDUCTIVITY

In a conductance cell (cell constant $= 456\cdot5 \text{ m}^{-1}$) the resistances of a 5×10^{-4} mol dm^{-3} aqueous solution of potassium chloride and a sample of the water with which this solution was made are $6\cdot13\times10^4$ Ω and 8×10^6 Ω, respectively, at 25°C. Calculate the molar conductivity of 5×10^{-4} mol dm^{-3} aqueous potassium chloride at this temperature.

The conductivity of the potassium chloride solution, allowing for that of the water, is given by

$$\varkappa(\text{KCl}) = \varkappa(\text{solution}) - \varkappa(\text{water})$$

$$= \left(\frac{456\cdot5}{6\cdot13\times10^4} - \frac{456\cdot5}{8\times10^6} \right) \Omega^{-1}\,\text{m}^{-1}$$

$$= 7\cdot39\times10^{-3}\,\Omega^{-1}\,\text{m}^{-1}$$

Molar conductivity, Λ, and conductivity, $\varkappa$, are related by the expression,

$$\Lambda = \frac{\varkappa}{c}$$

where c is the electrolyte concentration. Therefore,

$$\Lambda = \frac{7\cdot39\times10^{-3}}{5\times10^{-4}\times10^3}\,\frac{\Omega^{-1}\,\text{m}^{-1}}{\text{mol m}^{-3}}$$

$$= 1\cdot478\times10^{-2}\,\Omega^{-1}\,\text{m}^2\,\text{mol}^{-1}$$

EXAMPLE 7.2. LIMITING MOLAR CONDUCTIVITY OF A STRONG ELECTROLYTE

The molar conductivities, Λ, of aqueous sodium chloride at 25°C and at various concentrations, c, are given in the following table:

c/mol dm^{-3}	Λ/Ω^{-1} m^2 mol^{-1}
0·000 5	0·012 450
0·001	0·012 374
0·002	0·012 270
0·005	0·012 065
0·01	0·011 851

Plot a graph of Λ against $\sqrt{c}$ and determine the limiting molar conductivity at infinite dilution, Λ^∞. Compare the graph with the theoretical plot given by the Onsager equation. At 25°C the dielectric constant of water is 78·5 and the viscosity of water is $8·9 \times 10^{-4}$ kg m^{-1} s^{-1}.

Extrapolation of Λ versus $\sqrt{c}$ to zero concentration *(Fig. 7.1)* gives

$$\Lambda^\infty = 0·012\ 65 \ \Omega^{-1} \ \text{m}^2 \ \text{mol}^{-1}$$

For a 1–1 electrolyte, the Onsager limiting equation relating molar conductivity and concentration (both expressed in the above units) takes the form,

$$\Lambda = \Lambda^\infty - \left[\frac{8·24 \times 10^{-4}}{\eta(DT)^{1/2}} + \frac{8·204 \times 10^5}{(DT)^{3/2}} \Lambda^\infty \right] \sqrt{c}$$

where D is the dielectric constant, η the viscosity (in kg m^{-1} s^{-1}) and T the thermodynamic temperature.

Substituting appropriate numerical values

$$\Lambda = \Lambda^\infty - [(6·02 \times 10^{-3}) + (0·229 \times 0·012\ 65)] \sqrt{c}$$
$$= \Lambda^\infty - 8·92 \times 10^{-3} \sqrt{c}$$

It can be seen from the graph that the Onsager equation applies only to very dilute solutions.

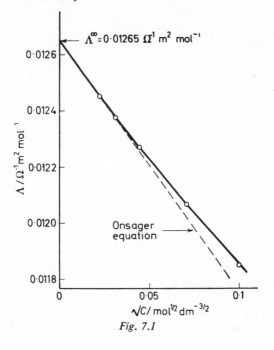

Fig. 7.1

EXAMPLE 7.3. LIMITING MOLAR CONDUCTIVITY OF A WEAK ELECTROLYTE

The limiting molar conductivities (in Ω^{-1} m^2 mol^{-1}) of aqueous sodium propionate, sodium chloride and hydrochloric acid are 0.859×10^{-2}, $1.264\ 5 \times 10^{-2}$ and $4.261\ 5 \times 10^{-2}$, respectively, at 25°C. Calculate the limiting molar conductivity of aqueous propionic acid at this temperature.

At very low concentrations each ion contributes a definite amount towards the total conductance of an electrolyte solution, irrespective of the nature of the other ions present.

$\varLambda^{\infty}$ for weak electrolytes cannot be measured directly, but may be calculated from the known limiting molar conductivities of the constituent ions,

e.g. $\varLambda^{\infty}(C_2H_5COOH) = \varLambda^{\infty}(H^+) + \varLambda^{\infty}(C_2H_5COO^-)$

From the data provided, $\varLambda^{\infty}$ for propionic acid is obtained as follows:

$$\Lambda^\infty(C_2H_5COOH) = \Lambda^\infty(C_2H_5COONa)+\Lambda^\infty(HCl)-\Lambda^\infty(NaCl)$$
$$= \Lambda^\infty(Na^+)+\Lambda^\infty(C_2H_5COO^-)+\Lambda^\infty(H^+)$$
$$+\Lambda^\infty(Cl^-)-\Lambda^\infty(Na^+)-\Lambda^\infty(Cl^-)$$
$$= \Lambda^\infty(H^+)+\Lambda^\infty(C_2H_5COO^-)$$

Substituting the appropriate numerical values,

$$\Lambda^\infty(C_2H_5COOH) = (0{\cdot}859+4{\cdot}261\ 5-1{\cdot}264\ 5)\times10^{-2}\ \Omega^{-1}m^2\ mol^{-1}$$
$$= 3{\cdot}856\times10^{-2}\ \Omega^{-1}\ m^2\ mol^{-1}$$

EXAMPLE 7.4. CONDUCTANCE AND IONIC MOBILITY

The conductivity of a $0{\cdot}01$ mol dm^{-3} aqueous solution of barium chloride at $25°C$ is $0{\cdot}238\ 2\ \Omega^{-1}\ m^{-1}$ and the transport number of the barium ions in this electrolyte is $0{\cdot}437\ 5$. Calculate the mobilities of the barium and chloride ions.

The molar conductivity of the barium chloride solution can be expressed as,

$$\Lambda(\tfrac{1}{2}BaCl_2) = \frac{\varkappa}{2c} = \frac{0{\cdot}238\ 2}{2\times0{\cdot}01\times10^3}\ \frac{\Omega^{-1}\ m^{-1}}{mol\ m^{-3}}$$
$$= 1{\cdot}191\times10^{-2}\ \Omega^{-1}\ m^2\ mol^{-1}$$

The molar conductivities of the barium and chloride ions are,

$$\Lambda(\tfrac{1}{2}Ba^{2+}) = \Lambda(\tfrac{1}{2}BaCl_2)\times t(Ba^{2+})$$
$$= 1{\cdot}191\times10^{-2}\times0{\cdot}437\ 5\ \Omega^{-1}\ m^2\ mol^{-1}$$
$$= 0{\cdot}521\times10^{-2}\ \Omega^{-1}\ m^2\ mol^{-1}$$

and
$$\Lambda(Cl^-) = \Lambda(\tfrac{1}{2}BaCl_2)\times t(Cl^-)$$
$$= 1{\cdot}191\times10^{-2}\times(1-0{\cdot}437\ 5)\ \Omega^{-1}\ m^2\ mol^{-1}$$
$$= 0{\cdot}670\times10^{-2}\ \Omega^{-1}\ m^2\ mol^{-1}$$

Ionic mobility, u, is given by the expression,

$$u = \frac{\Lambda}{F}$$

where F is Faraday's constant. Therefore,

$$u(Ba^{2+}) = \frac{0{\cdot}521\times10^{-2}}{96\ 500}\ \frac{\Omega^{-1}\ m^2\ mol^{-1}}{C\ mol^{-1}} = 5{\cdot}40\times10^{-8}\ m^2\ s^{-1}\ V^{-1}$$

and

$$u(Cl^-) = \frac{0.670 \times 10^{-2}}{96\,500} \frac{\Omega^{-1}\,m^2\,mol^{-1}}{C\,mol^{-1}} = 6.94 \times 10^{-8}\,m^2\,s^{-1}\,V^{-1}$$

EXAMPLE 7.5. TRANSPORT NUMBERS—HITTORF METHOD

In a Hittorf experiment aqueous silver nitrate solution was electrolysed between silver electrodes. The amount of silver nitrate in the anode compartment was 0·227 8 g before electrolysis and 0·281 8 g after electrolysis. During the electrolysis 0·019 4 g of copper was deposited on the cathode of a copper coulometer in series with the Hittorf cell. Calculate the transport numbers of the Ag^+ and NO_3^- ions.

The amount of Ag^+ in the anode compartment is

$$\frac{0.227\,8}{170}\,mol = 1.340 \times 10^{-3}\,mol \text{ (before electrolysis)}$$

and $\dfrac{0.281\,8}{170}\,mol = 1.658 \times 10^{-3}\,mol$ (after electrolysis)

i.e. the increase of Ag^+ during electrolysis is 0.318×10^{-3} mol.

If no Ag^+ ions had migrated out of the anode compartment this increase would have been $\dfrac{0.019\,4}{31.75}\,mol = 0.611 \times 10^{-3}$ mol, therefore, the amount of Ag^+ which has migrated is $(0.611 - 0.318) \times 10^{-3}\,mol = 0.293 \times 10^{-3}$ mol.

The transport number of the Ag^+ ions (i.e. the fraction of the total current carried by the Ag^+ ions) is, therefore, $\dfrac{0.293 \times 10^{-3}}{0.611 \times 10^{-3}}$

i.e. $\qquad t(Ag^+) = 0.480$

and $\qquad t(NO_3^-) = 1 - t(Ag^+) = 0.520$

EXAMPLE 7.6. TRANSPORT NUMBERS—MOVING BOUNDARY METHOD

In a tube of 8 mm diameter, the boundary between aqueous solutions of hydrochloric acid and sodium chloride moves with a velocity of 0·085 mm s^{-1} when the current is 5 mA. The concentration of the hydrochloric acid solution is 0·01 mol dm^{-3}. Calculate the transport number of the hydrogen ions.

In 1 second, 5×10^{-3} coulombs pass through the moving boundary cell and the number of moles of hydrogen ions transported across a cross-section of the cell equals $\dfrac{5 \times 10^{-3} \times t(H^+)}{96\ 500}$, where $t(H^+)$ is the transport number of the hydrogen ions.

This is equated with the number of moles of hydrogen ions transferred in 1 second across a cross-section of the cell, calculated from the velocity of the moving boundary, i.e.

$$\frac{5 \times 10^{-3} \times t(H^+)}{96\ 500} = 0.085 \times 10^{-3} \times \pi \times (4 \times 10^{-3})^2 \times 0.01 \times 10^3$$

giving $\qquad t(H^+) = 0.824$

EXAMPLE 7.7. CONDUCTANCE AND SOLUBILITY

At 298 K the conductivity of a saturated aqueous solution of silver chloride is $2.68 \times 10^{-4}\ \Omega^{-1}\ m^{-1}$ and that of the water with which the solution was made is $0.86 \times 10^{-4}\ \Omega^{-1}\ m^{-1}$. The limiting molar conductivities (in $\Omega^{-1}\ m^2\ mol^{-1}$) of aqueous silver nitrate, hydrochloric acid and nitric acid at 298 K are 1.33×10^{-2}, 4.26×10^{-2} and 4.21×10^{-2}, respectively. Calculate the solubility of silver chloride in water at this temperature.

The conductivity of the saturated solution is the sum of that due to the dissolved silver chloride and that of the solvent,

i.e. $\quad \varkappa(AgCl) = \varkappa(\text{solution}) - \varkappa(\text{water})$
$$= (2.68 - 0.86) \times 10^{-4}\ \Omega^{-1}\ m^{-1}$$
$$= 1.82 \times 10^{-4}\ \Omega^{-1}\ m^{-1}$$

The molar conductivity is given by

$$\Lambda(AgCl) = \frac{\varkappa(AgCl)}{c}$$

where c is the concentration (i.e. solubility) of silver chloride.

Since silver chloride is a strong electrolyte and only sparingly soluble, $\Lambda(AgCl)$ can be taken as equal to the limiting molar conductivity, which is given by

$$\Lambda^\infty(AgCl) = \Lambda^\infty(AgNO_3) + \Lambda^\infty(HCl) - \Lambda^\infty(HNO_3)$$
$$= (1.33 + 4.26 - 4.21) \times 10^{-2}\ \Omega^{-1}\ m^2\ mol^{-1}$$
$$= 1.38 \times 10^{-2}\ \Omega^{-1}\ m^2\ mol^{-1}$$

Therefore,

$$c = \frac{1 \cdot 82 \times 10^{-4}}{1 \cdot 38 \times 10^{-2}} \frac{\Omega^{-1}\,m^{-1}}{\Omega^{-1}\,m^2\,mol^{-1}} = 1 \cdot 32 \times 10^{-2}\,mol\,m^{-3}$$

$$= 1 \cdot 32 \times 10^{-5}\,mol\,dm^{-3}$$

$$= 1 \cdot 89 \times 10^{-3}\,g\,dm^{-3}$$

EXAMPLE 7.8. IONIC STRENGTH AND ACTIVITY COEFFICIENTS

Calculate the ionic strength of an aqueous solution of barium chloride at 298 K having a molality equal to $0 \cdot 002$ mol kg^{-1} and, using the Debye–Hückel limiting law, estimate (a) the activity coefficients of the Ba^{2+} and Cl^- ions in this solution, and (b) the mean ionic activity coefficients of these ions.

Ionic strength is defined by the expression

$$I = \tfrac{1}{2}\sum m_i z_i^2$$

where m_i and z_i are the molality and the charge number of ions of type i. Therefore, for $0 \cdot 002$ mol kg^{-1} BaCl$_2$

$$I = \tfrac{1}{2}[(1 \times 0 \cdot 002 \times 2^2) + (2 \times 0 \cdot 002 \times 1^2)]\,mol\,kg^{-1}$$

$$\quad\quad\quad (Ba^{2+}) \quad\quad\quad\quad (2Cl^-)$$

$$= 0 \cdot 006\,mol\,kg^{-1}$$

To estimate individual ionic activity coefficients, the Debye–Hückel limiting equation is used:

$$\log_{10} \gamma = -Az^2 \sqrt{I}$$

where $A = 0 \cdot 51$ kg$^{1/2}$ mol$^{-1/2}$ for aqueous solutions at 298 K, therefore,

$$\log_{10} \gamma_{Ba^{2+}} = -0 \cdot 51 \times 2^2 \times \sqrt{0 \cdot 006}$$

$$= -0 \cdot 158\,0 = \bar{1} \cdot 842\,0$$

giving $\quad\quad\quad \gamma_{Ba^{2+}} = 0 \cdot 695$

and $\quad\quad\quad \log_{10} \gamma_{Cl^-} = -0 \cdot 51 \times 1^2 \times \sqrt{0 \cdot 006}$

giving $\quad\quad\quad \gamma_{Cl^-} = 0 \cdot 913$

For an electrolyte which dissolves into ν_+ cations and ν_- anions, the mean ionic activity coefficient $\gamma_\pm$ is given by the expression

$$\gamma_\pm^\nu = \gamma_+^{\nu_+}\gamma_-^{\nu_-}$$

where $\nu = \nu_+ + \nu_-$; therefore, for $BaCl_2$,

$$\gamma_\pm^3 = \gamma_{Ba^{2+}}\gamma_{Cl^-}^2$$
$$= 0 \cdot 695 \times (0 \cdot 913)^2$$

giving $\qquad \gamma_\pm = 0 \cdot 834$

EXAMPLE 7.9. DISSOCIATION CONSTANT FROM CONDUCTANCE DATA (APPROXIMATE METHOD)

The molar conductivities, Λ, of aqueous acetic acid at 25°C and at various concentrations, c, are given in the following table:

$c/\text{mol dm}^{-3}$	$\Lambda/\Omega^{-1}\,\text{m}^2\,\text{mol}^{-1}$
0	0·039 07
0·000 4	0·007 39
0·000 9	0·005 115
0·002 5	0·003 165
0·01	0·001 621
0·04	0·000 819

Treat the data graphically to obtain an approximate value for the dissociation constant of acetic acid in water.

The dissociation of acetic acid and the equilibrium concentrations can be represented as follows:

$$HAc = H^+ + Ac^-$$

$$(1-\alpha)c \qquad \alpha c \qquad \alpha c$$

where α is the degree of ionization.

Sig. 8

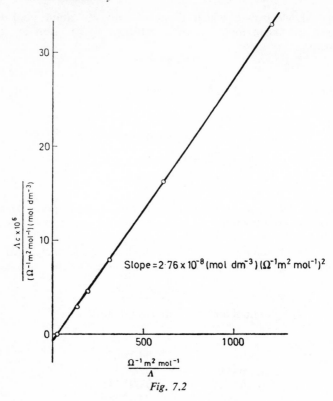

Fig. 7.2

The classical dissociation 'constant', K_a' (calculated in terms of concentrations), is given by

$$K_a' = \frac{c_{H^+}c_{Ac^-}}{c_{HAc}} = \frac{\alpha^2 c}{(1-\alpha)}$$

The degree of ionization of a weak electrolyte is approximately equal to the molar conductivity ratio, Λ/Λ^∞, therefore,

$$K_a' = \frac{\left(\dfrac{\Lambda}{\Lambda^\infty}\right)^2 c}{\left(1 - \dfrac{\Lambda}{\Lambda^\infty}\right)}$$

which rearranges to

$$\Lambda c = K_a'\left(\frac{(\Lambda^\infty)^2}{\Lambda} - \Lambda^\infty\right)$$

A plot of Λc against $1/\Lambda$ should, therefore, give a straight line of slope $K_a'(\Lambda^\infty)^2$.

$\dfrac{c}{\text{mol dm}^{-3}}$	$\dfrac{\Lambda \times c \times 10^6}{(\Omega^{-1}\,\text{m}^2\,\text{mol}^{-1})\,(\text{mol dm}^{-3})}$	$\dfrac{\Omega^{-1}\,\text{m}^2\,\text{mol}^{-1}}{\Lambda}$
0	0	26
0·000 4	2·96	135
0·000 9	4·60	195
0·002 5	7·91	316
0·01	16·2	617
0·04	32·8	1220

From the graph *(Fig. 7.2)*,

$$K_a' = \frac{\text{slope}}{(\Lambda^\infty)^2} = \frac{2\cdot76\times10^{-8}}{(0\cdot039\ 07)^2}\ \text{mol dm}^{-3}$$
$$= 1\cdot81\times10^{-5}\ \text{mol dm}^{-3}$$

K_a, the thermodynamic dissociation constant [calculated in terms of relative activities (relative to a standard concentration of 1 mol dm^{-3}, see page 14)] is, therefore, approximately $1\cdot81\times10^{-5}$.

Note. The errors involved in neglecting activity coefficients and in putting Λ/Λ^∞ equal to the degree of dissociation somewhat cancel one another and the final answer is close to the accepted value for K_a of $1\cdot754\times10^{-5}$.

EXAMPLE 7.10. pH OF A SOLUTION OF A WEAK ACID

At 25°C the molar conductivity of 0·01 mol dm^{-3} aqueous acetic acid is $1\cdot620\times10^{-3}\ \Omega^{-1}\ \text{m}^2\ \text{mol}^{-1}$ and the limiting molar conductivity at infinite dilution is $39\cdot07\times10^{-3}\ \Omega^{-1}\ \text{m}^2\ \text{mol}^{-1}$. Calculate (*a*) the pH of 0·01 mol dm^{-3} aqueous acetic acid at 25°C, and (*b*) the molar conductivity and pH of 0·1 mol dm^{-3} aqueous acetic acid at 25°C.

(*a*) The dissociation of acetic acid and the equilibrium concentrations can be represented as follows:

$$HAc = H^+ + Ac^-$$
$$(1-\alpha)c \quad \alpha c \quad \quad \alpha c$$

where c is the concentration of acetic acid and α the degree of ionization.

The degree of ionization of a dilute solution of a weak electrolyte is approximately equal to the molar conductivity ratio, Λ/Λ^∞. Therefore, for 0.01 mol dm^{-3} acetic acid,

$$\alpha = \frac{1.620 \times 10^{-3}}{39.07 \times 10^{-3}} = 0.041\ 5$$

and $\qquad c_{H^+} = \alpha c = 0.041\ 5 \times 0.01$ mol dm^{-3}

The ionic strength, αc, is sufficiently low for activity coefficients to be neglected with little error, therefore,

$$pH = -\log_{10}(c_{H^+}/\text{mol dm}^{-3}) = 3.38$$

(b) The classical dissociation function of acetic acid is given by,

$$K_a' = \frac{c_{H^+} c_{Ac^-}}{c_{HAc}} = \frac{\alpha^2 c}{(1-\alpha)}$$

$$= \frac{(0.041\ 5)^2 \times 0.01}{(1-0.041\ 5)} \text{ mol dm}^{-3} = 1.80 \times 10^{-5} \text{ mol dm}^{-3}$$

For 0.1 mol dm^{-3} acetic acid,

$$K_a' = \frac{\alpha^2 \times 0.1}{(1-\alpha)} \text{ mol dm}^{-3} = 1.80 \times 10^{-5} \text{ mol dm}^{-3}$$

giving $\qquad \alpha = 0.013\ 3$

Therefore, $\quad \Lambda(0.1 \text{ mol dm}^{-3}) = \alpha\Lambda^\infty$

$$= 0.013\ 3 \times 39.07 \times 10^{-3}\ \Omega^{-1}\text{ m}^2\text{ mol}^{-1}$$

$$= 0.520 \times 10^{-3}\ \Omega^{-1}\text{ m}^2\text{ mol}^{-1}$$

and, again neglecting the activity coefficient,

$$pH = -\log_{10}(c_{H^+}/\text{mol dm}^{-3}) = -\log_{10}(0.013\ 3 \times 0.1)$$

$$= 2.88$$

Example 7.11. pH of an Acetate–Acetic Acid Buffer Solution

An aqueous sodium acetate–acetic acid buffer solution of ionic strength 0.01 mol dm^{-3} and at 25°C is prepared by adding 0.02 mol dm^{-3} hydrochloric acid to 0.5 dm^3 of 0.02 mol dm^{-3}

sodium acetate and then making the volume up to 1.0 dm^3 with water. Calculate the volume of 0.02 mol dm^{-3} hydrochloric acid required to prepare a buffer solution of pH 5.0, (a) accurately, taking the activity coefficient of the acetate ions to be 0.90, and (b) approximately, using the Henderson equation. The thermodynamic dissociation constant of acetic acid is 1.75×10^{-5}.

(a) The final buffer solution contains undissociated HAc. plus Na$^+$, H$^+$, Ac$^-$, OH$^-$ and Cl$^-$ ions. For electroneutrality,

$$c_{Na^+} + c_{H^+} = c_{Ac^-} + c_{OH^-} + c_{Cl^-}$$

where c represents concentration.

$c_{Na^+} = 0.01$ mol dm^{-3}, and (if a volume, V, of 0.02 mol dm^{-3} hydrochloric acid is used to make up the buffer solution) $c_{Cl^-} = = 0.02\,(V/\text{dm}^3)$ mol dm^{-3}, therefore,

$c_{Ac^-} = 0.01$ mol dm$^{-3} - 0.02\,(V/\text{dm}^3)$ mol dm$^{-3} + c_{H^+} - c_{OH^-}$

The acetate ions of the sodium acetate either remain unchanged or they combine with the hydrogen ions of the added HCl to form undissociated HAc, therefore,

$$c_{HAc} = 0.01 \text{ mol dm}^{-3} - c_{Ac^-}$$
$$= 0.02\,(V/\text{dm}^3) \text{ mol dm}^{-3} - c_{H^+} + c_{OH^-}$$

The dissociation constant, K_a, of acetic acid is given by,

$$K_a = \frac{a_{H^+} a_{Ac^-}}{a_{HAc}} = \frac{a_{H^+} c_{Ac^-}}{c_{HAc}} \cdot \frac{y_{Ac^-}}{y_{HAc}}$$
$$= \frac{a_{H^+}(0.01 \text{ mol dm}^{-3} - 0.02\,(V/\text{dm}^3) \text{ mol dm}^{-3} + c_{H^+} - c_{OH^-})y_{Ac^-}}{(0.02\,(V/\text{dm}^3) \text{ mol dm}^{-3} - c_{H^+} + c_{OH^-})y_{HAc}}$$

where a represents relative activity and y activity coefficient. y_{HAc} is sufficiently close to unity to justify omission and c_{H^+} and c_{OH^-} (ca. 10^{-5} mol dm^{-3} and 10^{-9} mol dm^{-3}, respectively) can be neglected with little error, therefore,

$$K_a = a_{H^+}\left(\frac{0.01 - 0.02V/\text{dm}^3}{0.02V/\text{dm}^3}\right)y_{Ac^-}$$

i.e. $\qquad 1.75 \times 10^{-5} = 10^{-5} \times \left(\dfrac{0.01 - 0.02V/\text{dm}^3}{0.02V/\text{dm}^3}\right) \times 0.90$

giving, $\qquad\qquad V = 0.170$ dm^3 or 170 cm^3

(b) If y_{Ac^-} is taken to be unity, the above treatment leads to the Henderson equation, i.e.

$$K_a \approx a_{H^+} \frac{c_{salt}}{c_{acid}} \quad \left(\text{or, pH} \approx pK_a + \log_{10} \frac{c_{salt}}{c_{acid}} \right)$$

therefore, $1·75 \times 10^{-5} \approx 10^{-5} \times \left(\dfrac{0·01 - 0·02V/\text{dm}^3}{0·02V/\text{dm}^3} \right)$

giving, $V \approx 0·182 \text{ dm}^3$ or 182 cm^3

EXAMPLE 7.12. HYDROLYSIS CONSTANT FROM CONDUCTANCE DATA

Calculate the hydrolysis constant of aqueous aniline hydrochloride and the base dissociation constant of aniline at 25°C from the following conductance data:

0·01 mol dm^{-3} aniline
hydrochloride $\Lambda = 1·242 \times 10^{-2} \ \Omega^{-1} \text{ m}^2 \text{ mol}^{-1}$

0·01 mol dm^{-3}
hydrochloric acid $\Lambda = 4·120 \times 10^{-2} \ \Omega^{-1} \text{ m}^2 \text{ mol}^{-1}$

0·01 mol dm^{-3} aniline
hydrochloride +
0·01 mol dm^{-3} aniline $\Lambda = 1·091 \times 10^{-2} \ \Omega^{-1} \text{ m}^2 \text{ mol}^{-1}$

0·01 mol dm^{-3} aniline
hydrochloride +
0·02 mol dm^{-3} aniline $\Lambda = 1·090 \times 10^{-2} \ \Omega^{-1} \text{ m}^2 \text{ mol}^{-1}$

$K_w = 1·01 \times 10^{-14}$ at 25°C.

The hydrolysis of aniline hydrochloride and the equilibrium concentrations can be represented as follows:

$$C_6H_5NH_3^+ + H_2O = C_6H_5NH_2 + H_3O^+$$

$$(1-\alpha)c \qquad\qquad \alpha c \qquad \alpha c$$

where c represents the stoichiometric concentration of aniline hydrochloride (0·01 mol dm^{-3}) and α is the degree of dissociation.

The conductance of the aniline hydrochloride solution is due partly to the ions of the salt and partly to the ions produced by hydrolysis. If Λ, Λ' and Λ'' are the molar conductivities of 0·01 mol

dm^{-3} solutions of aniline hydrochloride, unhydrolysed aniline hydrochloride and hydrochloric acid respectively,

$$\Lambda = (1-\alpha)\Lambda' + \alpha\Lambda''$$

The molar conductivity, Λ', of unhydrolysed aniline hydrochloride is obtained from measurements on solutions in which hydrolysis has been suppressed by the addition of aniline. It is evident from the data that $0\cdot02$ mol dm^{-3} aniline is sufficient to suppress hydrolysis almost completely.

Substituting in the above equation

$$1\cdot242\times10^{-2} = 1\cdot090\times10^{-2}(1-\alpha) + 4\cdot120\times10^{-2}\alpha$$

giving $\qquad\qquad \alpha = 0\cdot050\ 2$

The hydrolysis constant is given by

$$K_h = \frac{a_{C_6H_5NH_2}a_{H_3O^+}}{a_{C_6H_5NH_3^+}} = \frac{c_{C_6H_5NH_2}c_{H_3O^+}}{c_{C_6H_5NH_3^+}c^\ominus}\cdot\frac{y_{C_6H_5NH_2}y_{H_3O^+}}{y_{C_6H_5NH_3^+}}$$

where a represents relative activity, y activity coefficient and $c^\ominus = 1$ mol dm^{-3} (see page 14). Since $y_{H_3O^+} \approx y_{C_6H_5NH_3^+}$ and $y_{C_6H_5NH_2} \approx 1$, then $\dfrac{y_{C_6H_5NH_2}\cdot y_{H_3O^+}}{y_{C_6H_5NH_3^+}} \approx 1$, and K_h can, therefore, be expressed with little error as

$$K_h = \frac{c_{C_6H_5NH_2}c_{H_3O^+}}{c_{C_6H_5NH_3^+}c^\ominus} = \frac{\alpha^2 c}{(1-\alpha)c^\ominus}$$

$$= \frac{(0\cdot050\ 2)^2\times0\cdot01}{(1-0\cdot050\ 2)} = 2\cdot65\times10^{-5}$$

The base dissociation constant, K_b, of aniline is given by

$$K_b = \frac{K_w}{K_h} = \frac{1\cdot01\times10^{-14}}{2\cdot65\times10^{-5}}$$

$$= 3\cdot81\times10^{-10}$$

EXAMPLE 7.13. pH OF A SOLUTION OF A SALT OF A WEAK ACID AND A STRONG BASE

At 25°C the dissociation constant, K_a, of acetic acid in water is $1\cdot75\times10^{-5}$ and the ionic product, K_w, of water is $1\cdot01\times10^{-14}$. Using the Debye–Hückel limiting law to estimate activity coefficients, calculate the pH of a $0\cdot01$ mol dm^{-3} aqueous solution of sodium acetate at this temperature.

The solution will be alkaline owing to the hydrolysis reaction

$$Ac^- + H_2O = HAc + OH^-$$
$$(1-\alpha)c \qquad \alpha c \qquad \alpha c$$

α is the degree of hydrolysis and c is the stoichiometric concentration of sodium acetate.

$$K_h \text{ (the hydrolysis constant)} = \frac{a_{HAc}a_{OH^-}}{a_{Ac^-}}$$

$$= \frac{c_{HAc}c_{OH^-}}{c_{Ac^-}c^\ominus} \cdot \frac{y_{HAc}y_{OH^-}}{y_{Ac^-}}$$

where a represents relative activity, y activity coefficient and $c^\ominus = 1$ mol dm^{-3} (see page 14). Since $y_{OH^-} \approx y_{Ac^-}$ and $y_{HAc} \approx 1$, then $\frac{y_{HAc}y_{OH^-}}{y_{Ac^-}} \approx 1$, and K_h can, therefore, be expressed with little error as

$$K_h = \frac{c_{HAc}c_{OH^-}}{c_{Ac^-}c^\ominus} = \frac{\alpha^2 c}{(1-\alpha)c^\ominus}$$

Also,
$$\frac{K_w}{K_a} = a_{H^+}a_{OH^-} \bigg/ \frac{a_{H^+}a_{Ac^-}}{a_{HAc}} = \frac{a_{HAc}a_{OH^-}}{a_{Ac^-}} = K_h$$

Therefore,
$$\frac{\alpha^2 \times 0.01}{(1-\alpha)} = \frac{1.01 \times 10^{-14}}{1.75 \times 10^{-5}}$$

giving
$$\alpha = 2.40 \times 10^{-4}$$

The hydrogen ion activity is given by

$$a_{H^+} = \frac{K_w}{a_{OH^-}} = \frac{K_w}{\alpha c y_{OH^-}}$$

Using the Debye–Hückel limiting law
$$\log_{10} y_{OH^-} = -0.51 \times \sqrt{0.01}$$

giving
$$y_{OH^-} = 0.89$$

Therefore,
$$pH = -\log_{10} a_{H^+}$$
$$= -\log_{10} \frac{1.01 \times 10^{-14}}{2.40 \times 10^{-4} \times 10^{-2} \times 0.89}$$
$$= 8.33$$

EXAMPLE 7.14. pH OF A SOLUTION OF A SALT OF A WEAK ACID AND A WEAK BASE

Calculate the pH of an aqueous solution of ethylammonium acetate at 25°C given the following equilibrium constants:

K_a (acetic acid) $= 1{\cdot}75 \times 10^{-5}$
K_b (ethylamine) $= 4{\cdot}7 \times 10^{-4}$
K_w $= 1{\cdot}01 \times 10^{-14}$

The hydrolysis of ethylammonium acetate and the equilibrium concentrations can be represented as follows:

$$EtNH_3^+ + Ac^- = EtNH_2 + HAc$$
$$(1-\alpha)c \quad (1-\alpha)c \quad \alpha c \quad \alpha c$$

where α is the degree of hydrolysis and c the stoichiometric concentration of ethylammonium acetate.

The hydrolysis constant is given by

$$K_h = \frac{a_{EtNH_2} a_{HAc}}{a_{EtNH_3^+} a_{Ac^-}} = \frac{c_{EtNH_2} c_{HAc}}{c_{EtNH_3^+} c_{Ac^-} y^2} = \frac{\alpha^2}{(1-\alpha)^2 y^2}$$

where a represents relative activity (see page 14) and c concentration. y is the activity coefficient of $EtNH_3^+$ and Ac^- ions (taken to be equal). The activity coefficients of $EtNH_2$ and HAc are taken to be unity.

Since $\quad \dfrac{K_w}{K_a K_b} = \dfrac{a_{H^+} a_{OH^-}}{\dfrac{a_{H^+} a_{Ac^-}}{a_{HAc}} \times \dfrac{a_{EtNH_3^+} a_{OH^-}}{a_{EtNH_2}}} = K_h$

then $\quad \dfrac{\alpha^2}{(1-\alpha)^2 y^2} = \dfrac{K_w}{K_a K_b}$

The hydrogen ion activity is given in terms of the dissociation constant of acetic acid by

$$a_{H^+} = K_a \frac{a_{HAc}}{a_{Ac^-}} = K_a \frac{c_{HAc}}{c_{Ac^-} y}$$

and, since $\quad \dfrac{c_{HAc}}{c_{Ac^-}} = \dfrac{\alpha}{(1-\alpha)},$

$$a_{H^+} = K_a \frac{\alpha}{(1-\alpha)y}$$

$$= K_a \left(\frac{K_w}{K_a K_b} \right)^{1/2} = \left(\frac{K_a K_w}{K_b} \right)^{1/2}$$

$$= \left(\frac{1 \cdot 75 \times 10^{-5} \times 1 \cdot 01 \times 10^{-14}}{4 \cdot 7 \times 10^{-4}} \right)^{1/2}$$

giving $\mathrm{pH} = -\log_{10} a_{H^+} = 7 \cdot 71$

EXAMPLE 7.15. ACID–BASE INDICATOR

A very small amount of phenolphthalein is added to a $0 \cdot 1$ mol dm^{-3} aqueous solution of sodium propionate at 25°C. Calculate the fraction of indicator which will assume the coloured form.

At 25°C: $K_w = 1 \cdot 01 \times 10^{-14}$
K_a (propionic acid) $= 1 \cdot 34 \times 10^{-5}$
K_{In} (phenolphthalein) $= 3 \cdot 16 \times 10^{-10}$

The indicator properties of phenolphthalein are a consequence of equilibria which can be represented as follows:

$$\underbrace{HIn_1}_{\text{colourless}} = \underbrace{HIn_2 = H^+ + In_2^-}_{\text{red}}$$

where HIn_1 and HIn_2 are tautomers, HIn_2 being ionizable.
The indicator constant, K_{In}, is defined by the expression

$$K_{In} = \frac{a_{H^+} a_{In_2^-}}{a_{HIn_1}}$$

where a represents relative activity. In this calculation it will be assumed that (a) all (univalent) ionized species have the same activity coefficient, y, (b) the activity coefficient of unionized species is unity, and (c) $c_{In_2^-} \gg c_{HIn_2}$ (where c represents concentration). Therefore,

$$K_{In} = \frac{a_{H^+} c_{(\text{red})} y}{c_{(\text{colourless})}}$$

$$= \frac{a_{H^+} y x}{(1-x)}$$

where x is the fraction of the phenolphthalein in the red form.

a_{H^+} is calculated as in Example 7.13, i.e.

$$K_h = \frac{K_w}{K_a} = \frac{1 \cdot 01 \times 10^{-14}}{1 \cdot 34 \times 10^{-5}} = \frac{\alpha^2 \times 0 \cdot 1}{1 - \alpha}$$

giving $\qquad\qquad \alpha = 8 \cdot 68 \times 10^{-5}$

where α is the degree of hydrolysis of the propionate ions,

and $\qquad a_{H^+} = \dfrac{K_w}{a_{OH^-}} = \dfrac{K_w}{y\alpha(c/\text{mol dm}^{-3})}$

$$= \frac{1 \cdot 01 \times 10^{-14}}{y \times 8 \cdot 68 \times 10^{-5} \times 0 \cdot 1}$$

Therefore,

$$3 \cdot 16 \times 10^{-10} = \frac{1 \cdot 01 \times 10^{-14} \times x}{8 \cdot 68 \times 10^{-5} \times 0 \cdot 1 \times (1 - x)} \qquad \text{(the activity coefficients cancel)}$$

giving $\qquad x = 0 \cdot 21$

EXAMPLE 7.16. HEAT OF NEUTRALIZATION

The ionic product of water as a function of temperature [Harned and Hamer, *J. Am. Chem. Soc.*, 55 (1933) 2194] is given in the following table:

$T/^\circ C$	0	10	25	40	50
$K_w \times 10^{14}$	0·113	0·292	1·008	2·917	5·474

Calculate ΔH for the ionization of water at 25°C.

K_w is the equilibrium constant for the reaction

$$2\,H_2O = H_3O^+ + OH^-$$

ΔH for this reaction can be calculated using the van't Hoff isochore, the integrated form of which is

$$\ln K_w = \frac{-\Delta H}{RT} + \text{constant}$$

i.e. $\qquad \Delta H = -2 \cdot 303 \times R \times (\text{slope of } \log_{10} K_w \text{ versus } 1/T)$

Therefore, $\Delta H_{25^\circ C} = -2 \cdot 303 \times 8 \cdot 314 \times (-2 \cdot 95 \times 10^3) \text{ J mol}^{-1}$

$\qquad\qquad\quad = 56 \cdot 5 \text{ kJ mol}^{-1}$

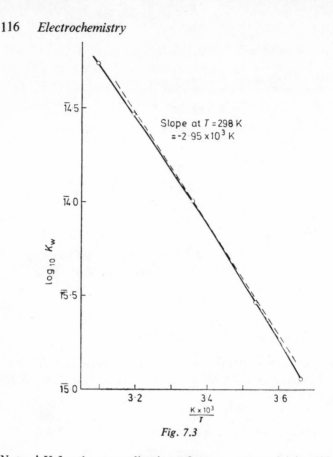

Slope at $T = 298$ K
$= -2.95 \times 10^3$ K

$\log_{10} K_w$

$\overline{14}.5$

$\overline{14}.0$

$\overline{15}.5$

$\overline{15}.0$

3.2   3.4   3.6

$\dfrac{K \times 10^3}{T}$

Fig. 7.3

Note. ΔH for the neutralization of a strong monobasic acid by a strong base in dilute aqueous solution is, therefore, equal to -56.5 kJ mol^{-1} at 25°C. The slight curvature of the graph *(Fig. 7.3)* suggests that ΔH is temperature-dependent.

EXAMPLE 7.17. NERNST EQUATION

Calculate the e.m.f. at 298 K of the cell

Zn$|$ZnSO$_4$(aq, 0.01 mol kg^{-1})$|\,|$KCl(aq, sat), Hg$_2$Cl$_2$(s)$|$Hg

The symbol $||$ represents a salt bridge, used to minimize liquid junction potential. At 298 K the standard potential of the zinc electrode is -0.763 V, the mean ionic activity coefficient, $\gamma_{\pm}$, of 0.01 mol kg^{-1} ZnSO$_4$ is 0.387 and the potential of the saturated calomel electrode is $+0.242$ V.

The e.m.f. of the cell is given by

$$E_{cell} = E_{calomel} - E_{Zn}$$

the calomel electrode being the positive pole.

The potential of the zinc electrode is calculated using the Nernst equation:

$$E = E^{\ominus} + \frac{RT}{zF} \ln \frac{a(\text{oxidized form})}{a(\text{reduced form})}$$

$$= E^{\ominus} + \frac{2 \cdot 303 RT}{zF} \log_{10} \frac{a_{Zn^{2+}}}{a_{Zn}}$$

where $E^{\ominus}$ is the standard electrode potential, F is Faraday's constant, z is the number of electrons involved in the electrode reaction and a represents relative activity (see page 13).

For a temperature of 298 K,

$$\frac{2 \cdot 303 RT}{F} = \frac{2 \cdot 303 \times 8 \cdot 31 \times 298}{96\,500} \frac{(\text{J K}^{-1} \text{ mol}^{-1})\,(\text{K})}{(\text{C mol}^{-1})}$$

$$= 0 \cdot 059\,1\ \text{V}$$

Assuming that $\gamma_{Zn^{2+}} = \gamma_{SO_4^{2-}}$, then $a_{Zn^{2+}} = 0 \cdot 01 \times 0 \cdot 387$ and, therefore, since $z = 2$ and $a_{Zn} = 1$,

$$E_{Zn} = -0 \cdot 763\ \text{V} + \left(\frac{0 \cdot 059\,1}{2} \log_{10} 0 \cdot 003\,87 \right)\ \text{V}$$

$$= -0 \cdot 834\ \text{V}$$

and $\quad E_{cell} = 0 \cdot 242\ \text{V} - (-0 \cdot 834)\ \text{V}$

$$= 1 \cdot 076\ \text{V}$$

EXAMPLE 7.18. CONCENTRATION CELLS

Calculate the e.m.f.s at 298 K of the following concentration cells:

(a) $\quad$ Ag | AgNO$_3$(aq, 0·01 mol kg^{-1}) ||
AgNO$_3$(aq, 0·1 mol kg^{-1} | Ag

(b) $\quad$ Ag | AgNO$_3$(aq, 0·01 mol kg^{-1}) ⫶
AgNO$_3$(aq, 0·1 mol kg^{-1} | Ag

(c) $\quad$ Ag | AgCl(s), NaCl(aq, 0·1 mol kg^{-1}) ||
NaCl(aq, 0·01 mol kg^{-1}), AgCl(s) | Ag

(d) $\quad$ Ag | AgCl(s), NaCl(aq, 0·1 mol kg^{-1}) ⫶
NaCl(aq, 0·01 mol kg^{-1}, AgCl(s) | Ag

given the following activity coefficients and transport numbers:

	$\gamma_\pm$	t_+
0·01 mol kg^{-1} AgNO$_3$	0·897	0·465
0·1 mol kg^{-1} AgNO$_3$	0·734	0·468
0·01 mol kg^{-1} NaCl	0·903	0·392
0·1 mol kg^{-1} NaCl	0·778	0·385

The potential of a silver electrode is given by

$$E_{Ag} = E^{\ominus}_{Ag} + \frac{RT}{F} \ln a_{Ag^+}$$

and that of a silver–silver chloride electrode by

$$E_{AgCl} = E^{\ominus}_{AgCl} - \frac{RT}{F} \ln a_{Cl^-}$$

The right hand electrode is the positive pole for each of the above cells.

(a) In this cell, liquid junction potential is virtually eliminated and the e.m.f. is given by

$$E_{cell} = \frac{RT}{F} \ln \frac{a_{Ag^+}(\text{in } 0\cdot1 \text{ mol kg}^{-1} \text{AgNO}_3)}{a_{Ag^+}(\text{in } 0\cdot01 \text{ mol kg}^{-1} \text{AgNO}_3)}$$

Assuming that $\gamma_{Ag^+} = \gamma_{NO_3^-}$

$$E_{cell} = \left(0\cdot059 \, 1 \times \log_{10} \frac{0\cdot1 \times 0\cdot734}{0\cdot01 \times 0\cdot897}\right) \text{ V}$$

$$= 0\cdot054 \text{ V}$$

(b) The e.m.f. of this cell includes a liquid junction potential and is given by

$$E_{cell} = 2t_- \frac{RT}{F} \ln \frac{a_{Ag^+}(\text{in } 0\cdot1 \text{ mol kg}^{-1} \text{AgNO}_3)}{a_{Ag^+}(\text{in } 0\cdot01 \text{ mol kg}^{-1} \text{AgNO}_3)}$$

where t_- is the transport number (assumed concentration-independent) of the NO$_3^-$ ions (i.e. the ions which are not in equilibrium with the electrodes). Assuming that $\gamma_{Ag^+} = \gamma_{NO_3^-}$

$$E_{cell} = \left(2 \times 0\cdot533 \, 5 \times 0\cdot059 \, 1 \times \log_{10} \frac{0\cdot1 \times 0\cdot734}{0\cdot01 \times 0\cdot897}\right) \text{ V}$$

$$= 0\cdot058 \text{ V}$$

(c) $E_{cell} = \dfrac{RT}{F} \ln \dfrac{a_{Cl^-}(\text{in } 0\cdot1 \text{ mol kg}^{-1} \text{NaCl})}{a_{Cl^-}(\text{in } 0\cdot01 \text{ mol kg}^{-1} \text{NaCl})}$

therefore, assuming that $\gamma_{Na^+} = \gamma_{Cl^-}$

$$E_{cell} = \left(0.059\,1 \times \log_{10} \frac{0.1 \times 0.778}{0.01 \times 0.903}\right) V$$

$$= 0.055 \text{ V}$$

(d) $\qquad E_{cell} = 2t_+ \frac{RT}{F} \ln \frac{a_{Cl^-}(\text{in } 0.1 \text{ mol kg}^{-1} \text{ NaCl})}{a_{Cl^-}(\text{in } 0.01 \text{ mol kg}^{-1} \text{ NaCl})}$

where t_+ is the transport number (assumed concentration-independent) of the Na^+ ions (again, the ions which are not in equilibrium with the electrodes), therefore, assuming that $\gamma_{Na^+} = \gamma_{Cl^-}$

$$E_{cell} = \left(2 \times 0.388\,5 \times 0.0591 \times \log_{10} \frac{0.1 \times 0.778}{0.01 \times 0.903}\right) V$$

$$= 0.043 \text{ V}$$

EXAMPLE 7.19. THERMODYNAMICS OF CELLS

The e.m.f. of the cell

$$Zn \,|\, ZnCl_2(0.05 \text{ mol dm}^{-3}) \,|\, AgCl \text{ (s), Ag}$$

is 1.015 V at 298 K, the silver electrode being positive, whilst the temperature coefficient of its e.m.f. is $-0.000\,492$ V K^{-1}. Write down the equation for the reaction occurring when the cell is allowed to discharge and calculate the changes in (a) free energy, (b) heat content (enthalpy), and (c) entropy attending this reaction at 298 K.

(G.R.I.C. Part 1, 1966)

$-$ve pole	$Zn - 2\,e \qquad \rightarrow Zn^{2+}$
$+$ve pole	$2\,AgCl(s) + 2\,e \rightarrow 2\,Ag + 2\,Cl^-$
Overall reaction	$Zn + 2\,AgCl(s) \rightarrow 2\,Ag + Zn^{2+} + 2\,Cl^-$

(a) The free energy change for the overall cell reaction is given by

$$\Delta G = -zEF$$

where E is the e.m.f. of the cell, F is Faraday's constant and z is the number of electrons involved in the overall reaction between 1 atom of Zn and 2 molecules of AgCl(s), i.e. $z = 2$. Therefore,

$$\Delta G_{298\,K} = -2 \times 1.015 \text{ V} \times 96\,500 \text{ C mol}^{-1}$$
$$= -196\,000 \text{ J mol}^{-1}$$
$$= -196 \text{ kJ mol}^{-1}$$

(b) The Gibbs–Helmholtz equation takes the form

$$\Delta G = \Delta H + T\left[\frac{\partial(\Delta G)}{\partial T}\right]_p$$

Therefore, substituting $-zEF$ for ΔG,

$$\Delta H = -zEF + zFT\left(\frac{\partial E}{\partial T}\right)_p$$

from which,

$$\Delta H_{298\ K} = -196\ \text{kJ mol}^{-1} - \frac{2\times96\ 500\times298\times0\cdot000\ 492}{1000}\ \text{kJ mol}^{-1}$$

$$= -224\ \text{kJ mol}^{-1}$$

(c) $$\Delta G = \Delta H - T\Delta S$$

i.e. $$\Delta S = \frac{\Delta H - \Delta G}{T}$$

$$= zF\left(\frac{\partial E}{\partial T}\right)_p$$

therefore, $\Delta S_{298K} = -2\times96\ 500\times0\cdot000\ 492\ \text{J K}^{-1}\ \text{mol}^{-1}$
$$= -95\cdot0\ \text{J K}^{-1}\ \text{mol}^{-1}$$

EXAMPLE 7.20. REDOX POTENTIALS

At 298 K the standard redox potentials of the electrodes, $Pt\,|\,Ce^{4+}$, Ce^{3+} and $Pt\,|\,Fe^{3+}$, Fe^{2+}, are 1·61 V and 0·77 V, respectively. What conclusion may be drawn from these observations?

In a solution containing both redox systems, the equilibrium
$$Ce^{4+} + Fe^{2+} = Ce^{3+} + Fe^{3+}$$

is established, and, at equilibrium, the potential of the Ce^{4+}, Ce^{3+} redox system equals that of the Fe^{3+}, Fe^{2+} redox system, i.e.

$$1\cdot61 + 0\cdot059\ \log_{10}\frac{a_{Ce^{4+}}}{a_{Ce^{3+}}} = 0\cdot77 + 0\cdot059\ \log_{10}\frac{a_{Fe^{3+}}}{a_{Fe^{2+}}}$$

giving $$\frac{a_{Ce^{3+}}a_{Fe^{3+}}}{a_{Ce^{4+}}a_{Fe^{2+}}} = 1\cdot7\times10^{14}$$

Since the equilibrium constant is so high, it is apparent that ferrous ions can be titrated quantitatively with ceric ions.

EXAMPLE 7.21. SOLUBILITY PRODUCT FROM ELECTRODE POTEN-
TIALS

The standard potentials of the electrodes $Ag|Ag^+(aq)$ and $Ag|$ $AgCl(s)$, $Cl^-(aq)$ are $+0.799\ 1$ V and $+0.222\ 4$ V, respectively at 298 K. Calculate the solubility product of silver chloride and the solubility of silver chloride in pure water at this temperature.

The potential of a silver–silver chloride electrode is given by

$$E_{AgCl} = E_{Ag}^{\ominus} + \frac{RT}{F}\ln a_{Ag^+}$$

where $E_{Ag}^{\ominus}$ is the standard potential of the $Ag|Ag^+$ electrode. Since K_{AgCl} (the solubility product of silver chloride) $= a_{Ag^+}a_{Cl^-}$,

$$E_{AgCl} = E_{Ag}^{\ominus} + \frac{RT}{F}\ln K_{AgCl} - \frac{RT}{F}\ln a_{Cl^-}$$

The standard potential, $E_{AgCl}^{\ominus}$, of the silver–silver chloride electrode refers to $a_{Cl^-} = 1$ (relative to a standard molality of 1 mol kg^{-1}), therefore,

$$E_{AgCl}^{\ominus} = E_{Ag}^{\ominus} + \frac{RT}{F}\ln K_{AgCl}$$

i.e. $0.222\ 4 = 0.799\ 1 + 0.059\ 15\ \log_{10} K_{AgCl}$

giving $K_{AgCl} = 1.79 \times 10^{-10}$

For a solution of silver chloride in pure water

$$a_{Ag^+} = a_{Cl^-} = \sqrt{K_{AgCl}}$$

therefore, $a_{Ag^+} = 1.34 \times 10^{-5}$

A saturated solution of AgCl in water is sufficiently dilute to justify putting activity coefficients equal to unity, therefore,
   solubility of AgCl in pure water $= 1.34 \times 10^{-5}$ mol kg^{-1}

Additional Examples

1. The resistance of an aqueous solution containing 0.624 g of $CuSO_4 \cdot 5H_2O$ per 100 cm^3 of solution in a conductance cell of cell constant 153.7 m^{-1} is 520 Ω at 25°C. Calculate the molar conductivity of this solution.

Sig. 9

2. The resistances of aqueous solutions of barium chloride at 25°C in a conductance cell with a cell constant of 150 m^{-1} are as follows:

Concentration /mol dm^{-3}	0·000 2	0·000 5	0·001	0·002
Resistance/Ω	27 520	11 160	5680	2905

Estimate the limiting molar conductivity at infinite dilution of aqueous barium chloride at 25°C.

3. The molar conductivities, Λ, of aqueous potassium chloride at 25°C and at various concentrations, c, are as follows:

c/mol dm^{-3}	0·000 5	0·001	0·005	0·01
Λ/Ω^{-1} m^2 mol$^{-1}\times$ 10^{-2}	1·478 1	1·469 5	1·435 5	1·412 7

Estimate the limiting molar conductivity at infinite dilution.

4. The limiting molar conductivities (in Ω^{-1} m^2 mol^{-1}) of aqueous ammonium chloride, sodium chloride and sodium hydroxide are $1·497\times10^{-2}$, $1·264\ 5\times10^{-2}$ and $2·478\times10^{-2}$, respectively, at 25°C. Calculate the limiting molar conductivity of aqueous ammonia at this temperature.

5. At 25°C the limiting molar conductivities at infinite dilution of potassium chloride and sodium nitrate are $1·498\ 5\times10^{-2}$ Ω^{-1} m^2 mol^{-1} and $1·215\ 9\times10^{-2}$ Ω^{-1} m^2 mol^{-1}, respectively. The transport numbers of potassium and sodium ions in these solutions are 0·490 6 and 0·412 4, respectively. Calculate the limiting molar conductivities of potassium nitrate and sodium chloride.

(G.R.I.C. Part 1, 1965)

6. The limiting molar conductivity of aqueous sodium chloride is $1·089\times10^{-2}$ Ω^{-1} m^2 mol^{-1} at 18°C and the transport number of the sodium ions in this electrolyte is 0·395. Calculate the limiting mobilities of sodium and chloride ions at this temperature.

[Liverpool Polytechnic, B.Sc.(1st year), 1966]

7. Λ^∞ (Na$_2$SO$_4$,aq) is equal to $2·602\times10^{-2}$ Ω^{-1} m^2 mol^{-1} at 25°C and the corresponding transport number of sodium ions is 0·385. Calculate the limiting mobilities of sodium and sulphate ions in aqueous solution at 25°C.

8. The limiting molar conductivities (in Ω^{-1} m^2 mol^{-1}) of some ions in aqueous solution at 25°C are,

$$H^+, \quad 349·8\times10^{-4}; \qquad Na^+, \quad 50·1\times10^{-4};$$
$$Cl^-, \quad 76·35\times10^{-4}; \qquad OH^-, \quad 198·3\times10^{-4}.$$

100 cm³ of 0·01 mol dm⁻³ aqueous hydrochloric acid is titrated with 0·01 mol dm⁻³ aqueous sodium hydroxide at this temperature. Calculate the conductivities after the addition of (*a*) 0, (*b*) 50 cm³, (*c*) 100 cm³ and (*d*) 150 cm³ of titrant. Explain any simplifying assumptions that are made in the calculation.

9. In a Hittorf experiment a 0·1 mol dm⁻³ aqueous solution o hydrochloric acid at 25°C was electrolysed between platinum electrodes. During the electrolysis 34·2 mg of copper was deposited on the cathode of a copper coulometer in series with the Hittorf cell. After electrolysis 50 cm³ of the hydrochloric acid solution (sufficient to incorporate all of the concentration change) was run off from the anode compartment, titrated, and found to have a concentration of 0·082 1 mol dm⁻³.

(*a*) Calculate the transport numbers of the hydrogen and chloride ions in this solution.

(*b*) If 50 cm³ of hydrochloric acid solution was run off from the cathode compartment, what would its concentration be?

10. In a Hittorf experiment an aqueous solution of potassium chloride at 25°C, containing 1·462 g of salt per 100 g of solution, was electrolysed between a silver anode and a silver–silver chloride cathode. During the electrolysis 0·323 6 g of silver was deposited on the cathode of a silver coulometer in series with the Hittorf cell. After electrolysis the anode solution (49·70 g) and the cathode solution (52·19 g) contained 1·244 g and 1·672 g of salt per 100 g of solution, respectively. Calculate the transport number of the potassium ions.

11. After electrolysis of a solution containing 3·69 g of AgNO₃ per kg of water, the anode compartment was analysed and found to contain 0·118 0 g of AgNO₃ in 23·14 g of water. In a Ag volta-meter in series with the electrolysis cell 0·039 0 g of Ag was deposited. Calculate the transport numbers of the silver and nitrate ions.
[University of Birmingham, B.Sc. (2nd year), 1967]

12. In a moving boundary experiment a vertically mounted tube contains (from the top) layers of aqueous solutions of lithium chloride, sodium chloride and sodium acetate. During the passage of current through the tube for a certain time (with the positive electrode at the top), the boundary between the lithium chloride and sodium chloride layers moves downwards 35 mm and that between the sodium chloride and sodium acetate layers moves upwards 55 mm. Calculate the transport numbers of the sodium and chloride ions in the sodium chloride solution.

13. The following data refer to a moving boundary experiment with 0·1 mol dm^{-3} potassium chloride using 0·065 mol dm^{-3} lithium chloride as indicator solution [MacInnes and Smith, *J. Am. Chem. Soc.*, 45 (1923) 2246]:

Current	5·893 mA
Cross-section of tube	11·42 mm^2
Boundary velocity	0·026 3 mm s^{-1}
Temperature	25°C

Given that the conductivity of 0·1 mol dm^{-3} potassium chloride at 25°C is 1·29 Ω^{-1} m^{-1}, calculate the transport number and the mobility of the potassium ions.

14. The conductivity at 25°C of a saturated aqueous solution of strontium sulphate is $1·48 \times 10^{-2}$ Ω^{-1} m^{-1} while that of the water from which the solution was made is $1·5 \times 10^{-4}$ Ω^{-1} m^{-1}. The limiting molar conductivities of $\frac{1}{2} Sr^{2+}$ and $\frac{1}{2} SO_4^{2-}$ are $5·95 \times 10^{-3}$ Ω^{-1} m^2 mol^{-1} and $8·00 \times 10^{-3}$ Ω^{-1} m^2 mol^{-1}, respectively. What is the solubility of strontium sulphate in water at 25°C?

[Liverpool Polytechnic, B.Sc. (1st year), 1968]

15. Calculate the solubility of lead sulphate in water at 18°C given that the conductivity of a saturated solution at this temperature is $1·65 \times 10^{-3}$ Ω^{-1} m^{-1} in excess of that of the water with which it is prepared and that the limiting mobilities of Pb^{2+} and SO_4^{2-} ions are $6·26 \times 10^{-8}$ m^2 s^{-1} V^{-1} and $7·10 \times 10^{-8}$ m^2 s^{-1} V^{-1}, respectively.

16. The conductivity of a solution of 0·001 mol dm^{-3} Na_2SO_4 is $2·6 \times 10^{-2}$ Ω^{-1} m^{-1}. Given that the molar conductivity of Na^+ is $5·0 \times 10^{-3}$ Ω^{-1} m^2 mol^{-1}, calculate the molar conductivity of the sulphate ion.

When this sodium sulphate solution is saturated with calcium sulphate the conductivity rises to $7·0 \times 10^{-2}$ Ω^{-1} m^{-1}. Calculate the solubility product of $CaSO_4$ given that the molar conductivity of $\frac{1}{2} Ca^{2+}$ is $6·0 \times 10^{-3}$ Ω^{-1} m^2 mol^{-1}.

[University of Leeds, B.Sc. (2CH), 1967]

17. Calculate, according to the Debye–Hückel limiting law, the activity coefficients of the Na^+, La^{3+}, Cl^- and NO_3^- ions in an aqueous solution at 298 K which is 0·002 mol kg^{-1} with respect to NaCl and 0·001 mol kg^{-1} with respect to $La(NO_3)_3$.

18. A weak monobasic acid is 5 per cent dissociated in 0·01 mol dm^{-3} solution. The limiting molar conductivity at infinite dilution

is 4.00×10^{-2} Ω^{-1} m^2 mol^{-1}. Calculate the conductivity of a 0·05 mol dm^{-3} solution of the acid.

19. The molar conductivity of nitric acid in methanol at 25°C is tabulated below as a function of the concentration:

c/mol dm^{-3} $\times 10^{-4}$	0	0·966	1·314	3·075	4·730	7·084	9·163
Λ/Ω^{-1} m^2 mol$^{-1} \times 10^{-4}$		203·0	178·7	165·2	151·0	137·6	124·2	116·3

Investigate the applicability of Ostwald's dilution law to this system.

(G.R.I.C., Part 1, 1968)

20. What is the conductivity at 25°C of distilled water in equilibrium with air at one atmosphere assuming that the air contains 0·050 volume per cent of CO_2? The only ions that need be considered are $H^+(aq)$ and $HCO_3^-(aq)$, their molar conductivities being 349.7×10^{-4} Ω^{-1} m^2 mol^{-1} and 44.5×10^{-4} Ω^{-1} m^2 mol^{-1}, respectively. At 25°C and at a partial pressure of 1 atm CO_2, 1 dm^3 of water dissolves 0·826 6 dm^3 CO_2. The first dissociation constant of carbonic acid, $H_2CO_3(aq)$, is 4.7×10^{-7} mol dm^{-3}.

(G.R.I.C., Part 1, 1968)

21. The degree of dissociation of chloroacetic acid in 0·01 mol dm^{-3} aqueous solution at 25°C is 0·33. Using the Debye–Hückel limiting law to estimate activity coefficients, calculate the thermodynamic dissociation constant of the acid.

22. The conductivity of a 0·05 mol dm^{-3} aqueous solution of acetic acid is 3.6×10^{-2} Ω^{-1} m^{-1} at 25°C. Given that at the same temperature the limiting mobilities at infinite dilution of hydrogen and acetate ions are 36.2×10^{-8} m^2 s^{-1} V^{-1} and 4.2×10^{-8} m^2 s^{-1} V^{-1}, respectively, calculate the pH of (a) 0·05 mol dm^{-3} and (b) 0·01 mol dm^{-3} solutions of acetic acid.

23. The pH of a 0·01 mol dm^{-3} aqueous solution of a weak monobasic acid is 3·3. Estimate the pH of a 0·2 mol dm^{-3} aqueous solution of this acid at the same temperature.

24. The molar conductivities at infinite dilution (in Ω^{-1} m^2 mol^{-1}) of aqueous sodium chloride, sodium formate and hydrochloric acid are 1.264×10^{-2}, 1.046×10^{-2} and 4.261×10^{-2}, respectively, at 25°C. The conductivity of 0·01 mol dm^{-3} aqueous formic acid at this temperature is 5.07×10^{-2} Ω^{-1} m^{-1}. Calculate (a) the limiting molar conductivity and the dissociation constant of

formic acid in aqueous solution at 25°C, and (b) the pH of 0·01 mol dm^{-3} aqueous formic acid at 25°C.

25. 25 cm^3 of 0·1 mol dm^{-3} formic acid was neutralized with sodium hydroxide solution, and a further 25 cm^3 of the acid added. The pH was measured and found to have a value of 3·82. Calculate (a) the pH of 0·1 mol dm^{-3} formic acid solution, (b) the pH of 0·005 mol dm^{-3} formic acid solution, and (c) the pH of 0·1 mol dm^{-3} formic acid solution which has been 75 per cent neutralized with sodium hydroxide solution.

[University of Salford, B.Sc. (2nd year), 1968]

26. 1·65 g of sodium acetate (anhydrous) is added to 500 cm^3 of 0·1 mol dm^{-3} aqueous acetic acid ($K_a = 1·8 \times 10^{-5}$). Calculate, approximately, (a) the pH of the resulting solution, and (b) the resulting pH if 10^{-3} mol of HCl is subsequently added to this solution.

27. Estimate the volume of 0·1 mol dm^{-3} sodium hydroxide which must be added to 200 cm^3 of 0·1 mol dm^{-3} acetic acid ($K_a = 1·75 \times 10^{-5}$) to give a solution with a pH of 6·0. Give a suitable choice of indicator for the titration of acetic acid with sodium hydroxide.

28. An aqueous buffer solution at 298 K, prepared by mixing 100 cm^3 of 0·1 mol dm^{-3} acetic acid with 100 cm^3 of 0·2 mol dm^{-3} sodium acetate, has a pH of 4·95. Using the extended Debye–Hückel equation, $\log_{10} y = \dfrac{-Az^2 \sqrt{I}}{1 + \sqrt{I}}$, to estimate activity coefficients, calculate the dissociation constant of acetic acid at 298 K.

29. The hydrolysis constant of aniline hydrochloride, $C_6H_5NH_3Cl$, is $2·63 \times 10^{-5}$ in water at 25°C. Calculate the degree of hydrolysis of a 0·1 mol dm^{-3} solution of aniline hydrochloride and the basic dissociation constant of aniline in water at 25°C. The ionic product of water at 25°C is $1·0 \times 10^{-14}$.

[Liverpool Polytechnic, B.Sc. (1st year), 1966]

30. Calculate the degree of hydrolysis and the pH of a 0·02 mol dm^{-3} aqueous solution of ammonium chloride at 25°C, given that the dissociation constant of ammonium hydroxide is $1·77 \times 10^{-5}$ and the ionic product of water is $1·01 \times 10^{-14}$. Use the Debye–Hückel limiting equation to estimate activity coefficients.

31. At 25°C the dissociation constant of benzoic acid in water is $6·3 \times 10^{-5}$ and the ionic product of water is $1·01 \times 10^{-14}$. Calculate the pH of (a) 0·01 mol dm^{-3}, and (b) 0·1 mol dm^{-3} aqueous solutions of sodium benzoate (i) taking all activity coeffici-

ents to be unity, and (*ii*) taking the activity coefficients of the monovalent ions to be 0·9 in 0·01 mol dm^{-3} solution and 0·8 in 0·1 mol dm^{-3} solution.

32. Calculate the ionic product of water at 25°C given that the pH of a 0·05 mol dm^{-3} aqueous solution of sodium acetate is 8·64 at this temperature and that the ionization constant of acetic acid is 1·75×10^{-5}. Use the extended form of the Debye–Hückel equation, $\log_{10} y = \dfrac{-Az^2\sqrt{I}}{1+\sqrt{I}}$, to estimate activity coefficients.

33. Calculate the pH of a 0·1 mol dm^{-3} aqueous solution of ammonium formate at 25°C given the following equilibrium constants:

$$K_a \text{ (formic acid)} = 1\cdot77\times10^{-4}$$
$$K_b \text{ (ammonium hydroxide)} = 1\cdot77\times10^{-5}$$
$$K_w = 1\cdot0\times10^{-14}$$

34. Given that K_{a2} for $H_3PO_4 = 8\cdot0\times10^{-8}$ at 25°C, calculate the pH of a 0·02 mol dm^{-3} solution of potassium dihydrogen phosphate, KH_2PO_4. Neglect the ionization of HPO_4^{2-}.
[University of Manchester, B.Sc. (1st exam), 1967]

35. If a very small amount of phenolphthalein is added to a 0·15 mol dm^{-3} solution of sodium benzoate, what fraction of the indicator will exist in the coloured form. State any assumptions that you make.

$$K_a \text{ (benzoic acid)} = 6\cdot2\times10^{-5}$$
$$K_w = 1\cdot01\times10^{-14}$$
$$K_{In} \text{ (phenolphthalein)} = 3\cdot16\times10^{-10}$$

[University of Salford, B.Sc. (Part 1), 1967]

36. The ionic product, K_w, of water is 0·681×10^{-14} at 20°C and 1·471×10^{-14} at 30°C. Assess the heat of neutralization of a strong monobasic acid by a strong base in dilute aqueous solution at 25°C.

37. The e.m.f. of the cell,

$$Hg\,|\,Hg_2Cl_2(s),\ KCl(aq,\ sat)\,|\,|\,KCl(aq,\ 0\cdot1\ mol\ kg^{-1}),\ AgCl(s)\,|\,Ag$$

is 0·047 V at 298 K. The potential of the saturated calomel electrode is 0·241 5 V and the standard potential of the silver–silver chloride electrode is 0·222 5 V. Estimate an activity coefficient for the chloride ions in 0·1 mol kg^{-1} aqueous potassium chloride.

38. Using the Debye–Hückel limiting equation to estimate activity coefficients, calculate the e.m.f. at 298 K of the concentration cell,

Pt; H_2(1 atm)|HCl(aq, 0·002 5 mol kg^{-1}), AgCl(s)|Ag|AgCl(s),

HCl(aq, 0·01 mol kg^{-1})|H_2(1 atm); Pt

39. The e.m.f. of the cell,

Pt; H_2(1 atm)|HCl(aq, 0·001 mol kg^{-1}), AgCl(s)|Ag|AgCl(s),

HCl(aq, 0·1 mol kg^{-1})|H_2(1 atm); Pt

is 0·227 V at 298 K. Estimate the mean ionic activity coefficient of 0·1 mol kg^{-1} HCl.

40. Calculate the transport number of the copper ions in dilute copper sulphate solution given that the e.m.f. at 298 K of the cell with transference

$(-)$ Cu|CuSO$_4$(aq, 0·001 mol kg^{-1})

CuSO$_4$(aq, 0·01 mol kg^{-1})|Cu $(+)$

is 0·0268 V. The mean ionic activity coefficients of 0·001 mol kg^{-1} and 0·01 mol kg^{-1} copper sulphate are 0·74 and 0·44 respectively.

41. A cell consists of two hydrogen electrodes dipping into the same 0·1 mol dm^{-3} solution of hydrochloric acid. One electrode is supplied with pure hydrogen at atmospheric pressure, the other with a mixture of hydrogen and argon also at atmospheric pressure. What is the mole fraction of hydrogen in this mixture when the e.m.f. of the cell is 10 mV at 300 K? In which direction would current flow in an external wire used to join the two electrodes?
(University of Leeds, B.Sc. 2T, 1967)

42. Calculate the change in standard free energy at 298 K for the reaction,

$$Zn + Cu^{2+} \rightarrow Zn^{2+} + Cu$$

given that the standard electrode potentials at 298 K for zinc and copper electrodes are $-0·761$ V and $+0·340$ V respectively.
(University of Nottingham, B.Sc. Part 1, 1966)

43. The standard potentials at 298 K for Sn^{2+}|Sn and Pb^{2+}|Pb are $-0·140$ V and $-0·126$ V, respectively. Calculate the ratio of Sn^{2+} to Pb^{2+} ion concentration when equilibrium is established at 298 K in the reaction

$$Sn(s) + Pb^{2+} = Sn^{2+} + Pb(s)$$

What is $\Delta G^{\ominus}_{298K}$ for this reaction?
(Liverpool Polytechnic, H.N.C., 1967)

44. For the cell,

Pt; H_2(1 atm)|HCl(aq, 0·01 mol dm^{-3})||

NaOH(aq, 0·01 mol dm^{-3})| H_2(1 atm); Pt

in which the liquid junction potential has been effectively eliminated by the use of a salt bridge, the e.m.f. is 0·585 V at 298 K. Show what information can be derived from this measurement. The mean ionic activity coefficients of hydrochloric acid and sodium hydroxide in 0·01 mol dm^{-3} aqueous solution are each 0·905 at 298 K.

(G.R.I.C., Part 1, 1964)

45. Calculate the solubility of AgBr in water at 298 K given the following standard electrode potentials:

$$Ag\,|\,Ag^+, \qquad E^{\ominus}_{298\ K} = +0·799\ V,$$
$$Ag\,|\,AgBr(s),\ Br^-, \qquad E^{\ominus}_{298\ K} = +0·073\ V.$$

CHAPTER 8

True-False and Multiple Choice Questions

TRUE-FALSE

Classify the following statements as true or false.

> In an examination, the usual marking scheme would be—
>
> | | |
> |---|---|
> | correct answer | 1 mark |
> | no answer | 0 mark |
> | incorrect answer | −1 mark |

1. In an ideal gas there would be no intermolecular forces.

2. Under the same conditions of temperature and pressure 1 mol of any gas occupies the same volume. (Leeds, 2T, 1968)

3. The virial coefficients for a gas are temperature-independent.

4. The average speed of a molecule in an ideal gas is proportional to the square root of the thermodynamic temperature.

5. In a mixture of nitrogen and oxygen gases at thermal equilibrium the average speed of the nitrogen molecules is the same as that of the oxygen molecules.

6. In a mixture of nitrogen and oxygen gases at thermal equilibrium the average translational energy of the nitrogen molecules is the same as that of the oxygen molecules.

7. At constant pressure the mean free path of a molecule in a gas increases with increasing temperature.

8. The mean free path of a molecule in a gas is independent of

130

temperature if the concentration of molecules is kept constant.
(Leeds, 2T, 1968)

9. The viscosities of gases decrease with increasing temperature.

10. At ordinary temperatures the heat capacities of diatomic gases increase with increasing temperature.

11. For all gases the molar heat capacity at constant pressure is greater than that at constant volume.

12. When a substance vaporizes, its entropy increases.

13. When a reaction between two liquids in a Dewar vessel is accompanied by a rise in temperature, ΔH for the reaction is positive.

14. The enthalpy of a system can only decrease when heat energy is transferred from the system to its surroundings.

15. The heat of formation of all chemical compounds is negative.
(Leeds, 2CH, 1967)

16. The change in enthalpy in going from one thermodynamic state to another is independent of the path taken.

17. ΔH for any reaction is independent of temperature.

18. Thermodynamics enables one to deduce the direction of a chemical change but not the rate of that change.

19. The second law of thermodynamics is another way of stating the law of conservation of entropy.   (Leeds, 2CH, 1967)

20. The second law of thermodynamics is only of statistical significance.

21. Entropy is an intensive property.

22. All reversible engines working between the same temperature limits are equally efficient.   (Leeds, 2T, 1967)

23. If a fixed mass of ideal gas is expanded from pressure p_1 to pressure p_2 the final volume is greater if the expansion is isothermal than if the expansion is adiabatic.   (Leeds, 2T, 1967)

24. When two ideal gases are mixed at constant temperature and pressure the entropy of the system increases.   (Leeds, 2CH, 1968)

25. All spontaneous reactions are exothermic.

26. At equilibrium the entropy of a system must be a maximum.
(Sheffield, 1st year, 1967)

27. At 273·15 K and atmospheric pressure, the molar entropy of $H_2O(l)$ is equal to that of $H_2O(s)$.

28. At 273·15 K and atmospheric pressure, the molar free energy of $H_2O(l)$ is equal to that of $H_2O(s)$.

29. At 273·15 K and atmospheric pressure, the chemical potential of $H_2O(l)$ is equal to that of $H_2O(s)$.

30. Chemical potential is an intensive property.

31. The equilibrium constant of an exothermic reaction decreases with increase in temperature.   (Manchester, 1st year, 1967)

32. The gas phase equilibrium, $2SO_2 + O_2 = 2SO_3$, shifts to the right if the pressure is increased.

33. $\left(\dfrac{\partial G}{\partial p}\right)_T = V$

34. Heats of solution are always negative.

35. The heat capacities of crystalline solids tend to approximately 25 J K^{-1} mol^{-1} at high temperatures.

36. In a one-component system there can never be more than one triple-point.   (Leeds, 2CH, 1968)

37. Liquid carbon dioxide is unstable at all temperatures at atmospheric pressure because it is below the triple-point pressure.

38. All binary mixtures of miscible liquids which show deviations from Raoult's law form an azeotropic solution at a particular concentration ratio.

39. The components of a binary mixture of miscible liquids can only be completely separated from one another by fractional distillation at a given pressure if the mixture obeys Raoult's law.

40. For an ideal solution the depression of freezing point is proportional to the molality of the solute.   (Leeds, 2CH, 1968)

41. The terms *activity coefficient* and *fugacity* are synonymous.

42. If a reaction shows first-order kinetics it must be a unimolecular reaction.

43. For a first-order reaction the half-life is independent of the initial concentration of reactant.   (Sheffield, 1st year, 1967)

44. The reaction, $2A = B + C$, occurs in a single step in the gas phase; if the pressure of A is doubled the rate of the reaction will increase by a factor of four.   (Sheffield, 1st year, 1967)

45. If the order of a gas phase reaction is 1·5, the mechanism must involve atoms or free radicals.

46. In general, the rates of chemical reactions approximately double for a 10 K rise in temperature at all temperatures.

47. The decomposition of a gas on a solid catalyst follows zero-order kinetics, which suggests that the gas is weakly adsorbed by the catalyst.

48. The rate of the photochemical process, $A + h\nu$ = products, is independent of the concentration of A.

49. The rate of decay of a radioactive species may be increased by heating it.   (Sheffield, 1st year, 1967)

50. The molar conductivity of an electrolyte decreases as the concentration is increased.

51. The transport number of sodium ions is the same in all aqueous solutions of sodium salts for a given temperature and concentration.

52. In aqueous solution the limiting molar conductivity of sodium ions is less than that of potassium ions because the sodium ions are more strongly hydrated.

53. Sodium acetate is only partially dissociated in aqueous solution.

54. Potassium chloride in 0.001 mol kg^{-1} aqueous solution is classified as a weak electrolyte because the mean ionic activity coefficient can be calculated accurately from the Debye–Hückel limiting equation.

55. A 10^{-8} mol kg^{-1} aqueous solution of HCl at 25°C has a pH of 8.

56. Since the solubility of silver iodide in water at 25°C is 10^{-8} mol kg^{-1}, its solubility product at this temperature is 10^{-16}.

57. The strength of an acid is independent of its solvent.

58. The potential of a standard hydrogen electrode is taken as zero at all temperatures.

59. In an electrochemical cell the standard Gibbs free energy decrease is greater than the work done if the cell is not operating reversibly. (Leeds, 2CH, 1968)

60. When the cell, $(-)$ Zn$|$ZnSO$_4$(aq)$||$CuSO$_4$(aq)$|$Cu $(+)$, discharges, the Cu electrode is the cathode.

61. In the absence of external forces all suspended particles, regardless of their size and shape, have the same average translational kinetic energy.

62. The adsorption of gases or vapours on to solid surfaces is always exothermic.

63. The adsorption of oxygen on to a certain solid surface occurs slowly at 400 K and more slowly at 350 K; the process is, therefore, more likely to be chemisorption than physical adsorption.

64. In general the spacings between molecular energy levels in gaseous molecules diminish in the order: electronic, vibrational, rotational, translational. (Leeds, 2CH, 1968)

65. Molecules vibrate at zero K. (Leeds, 2CH, 1967)

66. Molecules have rotational energy even at zero K. (Leeds, 2CH, 1968)

67. The energy separation between vibrational levels in a diatomic molecule is constant. (Leeds, 2CH, 1968)

68. Translational energy level spacing is independent of the total volume of the system. (Leeds, 2CH, 1968)

69. Rotational energy levels become more closely spaced the higher they are above the lowest energy level.

70. An electronic emission spectrum will contain more lines than the corresponding absorption spectrum.

MULTIPLE CHOICE

Select the correct answer (or answers) to the following questions.

> In an examination, a possible marking scheme would be—
>
> each choice of correct answer $\dfrac{1}{n}$ mark
>
> each incorrect choice $\dfrac{-1}{m-n}$ mark
>
> where m is the total number of choices and n the number of these which are correct answers.

1. The volume occupied by 1 mol of helium at 10^5 N m^{-2} and 1000 K is approximately: (a) 22·4 dm^3; (b) 83 dm^3; (c)108 dm^3; (d) 770 dm^3; (e) 0·82 m^3.

2. A suitable equation of state for a real gas at high temperatures and pressures (which allows for the space occupied by the molecules by a volume factor, b) is; (a) $pV = bRT$; (b) $pV = RT+b$; (c) $pV = RT-b$; (d) $pV = RT+bp$; (e) $pV = RT-bp$.

3. If a gas absorbs 200 J of heat and expands by 500 cm^3 against a constant pressure of 2×10^5 N m^{-2}, then the change in thermodynamic (internal) energy is: (a) -300 J; (b) -100 J; (c)$+100$ J; (d) $+300$ J.

4. When a gas expands adiabatically against a finite pressure: (a) its internal energy always decreases; (b) the temperature always decreases; (c) its entropy always remains constant.

5. The translational energy of 1 mol of ideal gas at 25°C is approximately: (a) 600 J; (b) 1250 J; (c) 2500 J; (d) 3750 J.

6. Three molecules have velocities 100 m s^{-1}, 200 m s^{-1} and 300 m s^{-1}, respectively. The root mean square velocity is: (a) 190 m s^{-1}; (b) 200 m s^{-1}; (c) 216 m s^{-1}; (d) 400 m s^{-1}; (e) 467 m s^{-1}.

7. The effect of the thermodynamic temperature, T, on gaseous molecular collision rates is: (a) $Z \propto T^{1/2}$; (b) $Z \propto T$; (c) $Z \propto \exp(-\text{constant}/T)$; (d) Z independent of T.

8. Which of the following relationships are based on the assumption of ideal gaseous behaviour: (a) $\Delta U = q-w$; (b) $\dfrac{d \ln p}{dT} = \dfrac{\Delta H}{RT^2}$; (c) for a reversible adiabatic expansion or compression, $pV^\gamma = \text{constant}$; (d) $C_p-C_V=R$; (e) at constant pressure, $\Delta H = \Delta U+p\,\Delta V$; (f) $pV = \frac{1}{3}nmc^2$.

9. Given the following heats of formation:

$$H_2O(g), \quad \Delta H_{f,298K} = -242 \text{ kJ mol}^{-1}$$

$$CO(g), \quad \Delta H_{f,298K} = -111 \text{ kJ mol}^{-1}$$

ΔH_{298K} for the reaction

$$H_2O(g) + C(s) = H_2(g) + CO(g)$$

is: (a) -353 kJ mol^{-1}; (b) -131 kJ mol^{-1}; (c) $+131$ kJ mol^{-1}; (d) $+353$ kJ mol^{-1}; (e) more information is required.

10. For any process: (a) the heat absorbed by the system is independent of the reaction path; (b) the work done by the system is independent of the reaction path; (c) the change in the internal energy of the system is independent of the reaction path.

11. The bond energy of a C—H bond in methane refers to: (a) ΔH for the reaction, $CH_4(g) = C(s) + 2H_2(g)$; (b) ΔH for the reaction, $CH_4(g) = C(g) + 4H(g)$; (c) $\frac{1}{4}$ of ΔH for reaction (a); (d) $\frac{1}{4}$ of ΔH for reaction (b); (e) none of these.

12. For an ideal gas: (a) $\left(\frac{\partial U}{\partial T}\right)_V = 0$; (b) $\left(\frac{\partial U}{\partial V}\right)_T = 0$; (c) $\left(\frac{\partial U}{\partial T}\right)_p = 0$; (d) $\left(\frac{\partial U}{\partial p}\right)_T = 0$.

13. The total entropy of an isolated system in which a change takes place at a finite rate: (a) always increases; (b) always decreases; (c) remains constant; (d) may increase or decrease.

14. The maximum efficiency of a heat engine working between 100°C and 25°C is: (a) 100%; (b) 75%; (c) 25%; (d) 20%.

15. For the reaction, $CO(g) + H_2O(g) = H_2(g) + CO_2(g)$: (a) K_p is unity; (b) $K_p = K_c$; (c) $K_p > K_c$; (d) $K_p < K_c$; (e) K_p can have units of N m^{-2}.

16. For most liquids the molar entropy of vaporization at the normal boiling point is: (a) 8·3 J K^{-1} mol^{-1}; (b) 20 J K^{-1} mol^{-1} (c) 25 J K^{-1} mol^{-1}; (d) 85 J K^{-1} mol^{-1}; (e) 175 J K^{-1} mol^{-1}.

17. A solution in which the solvent obeys Raoult's law and the solute obeys Henry's law is: (a) a perfectly ideal solution; (b) an ideal dilute solution; (c) neither of these.

18. For an ideal solution: (a) $\Delta H_{mix} = 0$; (b) $\Delta S_{mix} = 0$; (c) $\Delta G_{mix} = 0$; (d) $\Delta V_{mix} = 0$.

19. A binary mixture which forms an ideal solution: (a) has only very weak interaction between solute and solvent molecules; (b) can be separated into its two components by repeated distillation; (c) has a vapour pressure intermediate between the vapour pressures of the pure components.

20. At 400 K liquid A has a vapour pressure of 4×10^4 N m^{-2} and liquid B a vapour pressure of 6×10^4 N m^{-2}. A and B form

an ideal solution. The mole fraction of B in the vapour which is in equilibrium with a solution containing 0·6 mole fraction of A is: (a) 0·31; (b) 0·40; (c) 0·50; (d) 0·60; (e) 0·69.

21. At atmospheric pressure, an azeotropic solution: (a) cannot be separated into its components by fractional distillation; (b) can be separated into its components by fractional distillation; (c) can be separated into its components by a single distillation.

22. The vapour pressure of a liquid, A, in the presence of a second liquid, B, with which it is immiscible is: (a) proportional to the mole fraction of A in the system; (b) independent of the mole fraction of A in the system; (c) a logarithmic function of temperature.   (Manchester, 1st year, 1967)

23. The half-life of a second order process, $2A = $ products, is: (a) independent of the initial concentration of A; (b) directly proportional to the initial concentration of A; (c) inversely proportional to the initial concentration of A.

24. A second order rate constant can have the units: (a) dm^3 mol^{-1} min^{-1}; (b) cm^3 $molecule^{-1}$ s^{-1}; (c) mol m^{-3} s^{-1}; (d) $molecule$ cm^{-3} s^{-1}; (e) mol^2 m^{-6} s^{-1}; (f) m^2 N^{-1} s^{-1}.

25. The hydrolysis of an ester in the presence of dilute HCl follows first order kinetics because: (a) the acid acts as a catalyst; (b) the rate is independent of the hydrogen ion concentration; (c) the hydrogen ion concentration is essentially constant throughout the reaction.

26. If ΔH for a reaction is $+100$ kJ mol^{-1}, the activation energy: (a) must be equal to or less than 100 kJ mol^{-1}; (b) must be equal to or greater than 100 kJ mol^{-1}; (c) may be greater or less than 100 kJ mol^{-1}.

27. If ΔH for a reaction is -100 kJ mol^{-1}, the activation energy: (a) must be equal to or less than 100 kJ mol^{-1}; (b) must be equal to or greater than 100 kJ mol^{-1}; (c) may be greater or less than 100 kJ mol^{-1}.

28. The mobility of an ion can have the unit: (a) m s^{-1}; (b) m s^{-1} V^{-1}; (c) m^2 s^{-1} V^{-1}.

29. The ionic strength of 0·1 mol kg^{-1} aqueous barium chloride is: (a) 0·1 mol kg^{-1}; (b) 0·15 mol kg^{-1}; (c) 0·2 mol kg^{-1}; (d) 0·3 mol kg^{-1}; (e) 0·4 mol kg^{-1}.

30. The pH of 0·1 mol dm^{-3} aqueous HCN at 25°C ($K_a = 10^{-9}$) is: (a) 4·0; (b) 5·0; (c) 9·0; (d) 10·0.

31. A solution of a weak monobasic acid and one of its salts will have maximum buffer efficiency when: (a) the concentrations of acid and salt are equal; (b) the pH is 7; (c) the pH is equal to the pH of a solution of the salt; (d) the pH is equal to the pK_a of the acid.

32. The pH of an aqueous solution at 25°C made up to be 0·1 mol dm^{-3} with respect to NaOH and 0·3 mol dm^{-3} with respect to acetic acid (pK_a = 4·75) would be approximately: (a) 4·25; (b) 4·45; (c) 4·75; (d) 5·05; (e) 5·25.

33. At a certain temperature, the solubility of Ca(IO$_3$)$_2$ is 2×10^{-3} mol dm^{-3}; its solubility product is, therefore: (a) 2×10^{-3} (b) 4×10^{-6}; (c) 8×10^{-9}; (d) 3·2×10^{-8}.

(Sheffield, 1st year, 1967)

34. In which of the following cells will the e.m.f. be independent of the activity of the chloride ions: (a) Zn|ZnCl$_2$(aq)|Cl$_2$; Pt; (b) Zn|ZnCl$_2$(aq)||KCl(aq), AgCl(s)|Ag; (c) Ag|AgCl(s), KCl(aq)|Cl$_2$; Pt; (d) Hg|Hg$_2$Cl$_2$(s), KCl(aq)||AgNO$_3$(aq)|Ag; (e) Pt; H$_2$|HCl(aq)|Cl$_2$; Pt.

35. The Langmuir adsorption isotherm is based on the assumption of: (a) a constant heat of adsorption; (b) a flat solid surface; (c) ideal gaseous behaviour; (d) zero activation energy for adsorption and desorption.

36. According to the Lambert–Beer law, the intensity of light transmitted through a medium of concentration, c: (a) increases linearly with c; (b) increases exponentially with c; (c) falls off linearly with c; (d) falls off exponentially with c.

37. At room temperature most molecules are in the (a) ground electronic state, (b) ground vibrational state, (c) ground rotational state.

38. Radiation of wavelength 600 nm corresponds to an energy of approximately: (a) 42 kJ mol^{-1}; (b) 112 kJ mol^{-1}; (c) 200 kJ mol^{-1}; (d) 500 kJ mol^{-1}.

Answers to additional examples

Chapter 2. Gases

1. (a) 0.8×10^5 N m^{-2}; (b) $p(N_2) = 0.32 \times 10^5$ N m^{-2}, $p(O_2) = 0.48 \times 10^5$ N m^{-2}; (c) $x(N_2) = 0.4$, $x(O_2) = 0.6$
2. (van der Waals constants: $a = 0.136\ 5$ N m^4 mol^{-2}, $b = 38.6 \times 10^{-6}$ m^3 mol^{-1}) $p = 22.25 \times 10^5$ N m^{-2}
3. (van der Waals constants: $a = 0.423\ 5$ N m^4 mol^{-2}, $b = 37.3 \times 10^{-6}$ m^3 mol^{-1}) Volume (s.t.p.) $= 0.022\ 3$ m^3 mol^{-1}
4. (a) $0.044\ 1$ mol; (b) 1.34×10^{-10} m ($b = 24.0 \times 10^{-6}$ m^3 mol^{-1})

5. $M = RT \lim_{p \to 0} \dfrac{\rho}{p} = 17.03 \times 10^{-3}$ kg mol^{-1}, i.e. $M_r = 17.03$
6. w (min) (reversible compression) $= 4000$ J mol^{-1}; w (max) (instantaneous compression) $= 9920$ J mol^{-1} (compare with corresponding answers in worked example 2.2)
7. $w = 10/64\ RT = 380$ J   8. 3.0×10^5 N m^{-2}
9. T_2 (max) (instantaneous adiabatic compression) $= 970$ K
10. $w = 2230$ J mol^{-1} (for both H_2 and ideal gas)
11. 31.38 kJ mol^{-1}
12. A; C_V(max) $= 12.5$ J K^{-1} mol^{-1}
    O_2; C_V(max) $= 29$ J K^{-1} mol^{-1}
    H_2S; C_V(max) $= 50$ J K^{-1} mol^{-1}
13. 643 K   14. 0.12
15. (a) $n = 3.22 \times 10^{10}$ molecules cm^{-3}; (b) $\lambda = 54$ m; (c) $\bar{t} = 0.121$ s
16. $Z = 1.02 \times 10^{35}$ m^{-3} s^{-1}; $\lambda = 5.96 \times 10^{-8}$ m; $\bar{c}(g) \approx \bar{c}(l)$, $\lambda(g) \gg \lambda(l)$
17. $\lambda = 4.31 \times 10^{-8}$ m; $\sigma = 4.41 \times 10^{-10}$ m; $\eta \propto T^{1/2}$ and independent of pressure
18. $\sigma = 5.16 \times 10^{-10}$ m   19. 22% CO, 78% CO_2
20. 2.4×10^{25} molecules, 1760 g

Chapter 3. Thermochemistry

1. $\Delta H^{\ominus}_{298\,K}$(formation of C_2H_2) $\qquad = +228$ kJ mol^{-1}
   $\Delta H^{\ominus}_{298\,K}$(formation of C_2H_4) $\qquad = +\ 51$ kJ mol^{-1}
   $\Delta H^{\ominus}_{298\,K}$(formation of C_2H_6) $\qquad = -\ 84$ kJ mol^{-1}
   $\Delta H^{\ominus}_{298\,K}$(hydrogenation of C_2H_2 to C_2H_6) $= -312$ kJ mol^{-1}
   $\Delta H^{\ominus}_{298\,K}$(hydrogenation of C_2H_4 to C_2H_6) $= -135$ kJ mol^{-1}
2. $\Delta H = -132 \cdot 8$ kJ mol^{-1} $\qquad$ 3. 429 kJ
4. $\Delta U_{298\,K}$(formation of CO) $= -111 \cdot 7$ kJ mol^{-1}
   $\Delta U_{298\,K}$(formation of CO_2) $= -393 \cdot 5$ kJ mol^{-1}
5. $\Delta H^{\ominus}_{298\,K}$(formation of benzoic acid) $= -387$ kJ mol^{-1}
6. $\Delta H^{\ominus} = -103$ kJ mol^{-1}
7. $\Delta H^{\ominus}_{298\,K}$[formation of NaOH(aq)] $= -470$ kJ mol^{-1}
   $\left(\Delta H^{\ominus}_{298\,K}\text{[formation of }H^+(aq)] = 0\right)$
8. $\Delta H_{298\,K}$(formation of N_2O) $= +82 \cdot 4$ kJ mol^{-1}
   $\Delta H_{423\,K}$(formation of N_2O) $= +82 \cdot 2$ kJ mol^{-1}
9. $\Delta H_{398\,K} = -48 \cdot 4$ kJ mol^{-1}; $\Delta U_{398\,K} = -45 \cdot 1$ kJ mol^{-1}
10. $\Delta H_{1000K} = -30 \cdot 7$ kJ mol^{-1}
11. $D(CH_3-Cl) = 325$ kJ mol^{-1}
12. $\Delta H_f^{\ominus}(HS^*) = +138 \cdot 5$ kJ mol^{-1}; $D(HS^*) = +356 \cdot 5$ kJ mol^{-1}
13. 287 kJ mol^{-1}
14. C_3H_6, strain energy $= 114$ kJ mol^{-1}; C_5H_{10}, strain energy $= 25$ kJ mol^{-1}, i.e. C_5H_{10} is more stable than C_3H_6
15. $A = 335$ kJ mol^{-1}

Chapter 4. Thermodynamics and Chemical Equilibria

1. $\eta_{max} = 0 \cdot 242$; $w = 33 \cdot 0$ kJ; $q = 25 \cdot 0$ kJ; $T_1 = 358 \cdot 5$ K
2. $\Delta S = 10 \cdot 2$ J K^{-1} $\qquad$ 3. $\Delta S = 8 \cdot 57$ J K^{-1} mol^{-1}
4. (a) $\Delta H = 4010$ J mol^{-1}, $\Delta S = 9 \cdot 55$ J K^{-1} mol^{-1}
   (b) $\Delta H = 0$, $\Delta S = -19 \cdot 15$ J K^{-1} mol^{-1}
5. $\Delta S = 187 \cdot 7$ J K^{-1} mol^{-1}
6. $w = 3 \cdot 1$ kJ mol^{-1}, $\Delta U = 36 \cdot 9$ kJ mol^{-1}, $\Delta H = 40 \cdot 0$ kJ mol^{-1},
   $\Delta G = 0$, $\Delta S = 107$ J K^{-1} mol^{-1}, $\Delta A = -3 \cdot 1$ kJ mol^{-1}
7. $2\,C + 2\,H_2 = C_2H_4$, $\qquad \Delta H^{\ominus}_{298\,K} = +33 \cdot 0$ kJ mol^{-1},
   $\qquad \Delta G^{\ominus}_{298\,K} = +49 \cdot 1$ kJ mol^{-1} $\qquad\qquad\qquad\qquad$ (1)
   $2\,C + 3\,H_2 = C_2H_6$, $\qquad \Delta H^{\ominus}_{298\,K} = -86 \cdot 0$ kJ mol^{-1},
   $\qquad \Delta G^{\ominus}_{298\,K} = -34 \cdot 4$ kJ mol^{-1} $\qquad\qquad\qquad\qquad$ (2)
   Since $\Delta G^{\ominus}$ for process (1) is positive and $\Delta G^{\ominus}$ for process (2) is negative, process (2) is thermodynamically more feasible.
8. (a) $C_2H_2 + H_2 = C_2H_4$, $\qquad \Delta G^{\ominus}_{298\,K} = -141$ kJ mol^{-1},
   $\qquad \Delta H^{\ominus}_{298\,K} = -174$ kJ mol^{-1},

(b) $C_2H_6 = C_2H_4 + H_2$, $\Delta G^{\ominus}_{298\,K} = 101$ kJ mol^{-1},
$\Delta H^{\ominus}_{298\,K} = 137$ kJ mol^{-1}
Therefore, process (a) is a feasible method of synthesis, whereas process (b) is not. Process (a) is favoured by an increase in pressure and a decrease in temperature. Process (b) is favoured by a decrease in pressure and an increase in temperature.

9. $K_p = 6.76 \times 10^{-5}$ atm; $K_c = 1.80 \times 10^{-6}$ mol dm^{-3}
10. n-C_5H_{12} : iso-C_5H_{12} : neo-C_5H_{12} = 1 : 2.64 : 0.37
                        = 0.249 : 0.659 : 0.092
11. $K_c = 3.92$ and 4.00; $\Delta G^{\ominus}_{282\,K} = -3.23$ kJ mol^{-1}
12. $\Delta G_{298\,K} = 17.1$ kJ mol^{-1}. If, in addition, the mean enthalpy change over the range of temperature 500 K to T_2 K is known, $K_p(T_2$ K), and, therefore, $\alpha(T_2$ K), can be obtained from the van't Hoff isochore.
13. $\Delta G^{\ominus}_{333\,K} = -810$ J mol^{-1}; $\Delta G^{\ominus}_{373\,K} = -5872$ J mol^{-1}; $\Delta H^{\ominus}$ = 41.3 kJ mol^{-1}
14. $K_p(2200$ K$) = 6.08 \times 10^{-5}$ atm; $\quad /K_p(2500$ K$) = 1.04 \times 10^{-3}$ atm; $\Delta H^{\ominus}_{\text{dissociation}} = 432$ kJ mol^{-1}
15. $\Delta H^{\ominus} = 15.8$ kJ mol^{-1}
16. $K_p(1573$ K$) = 2.90$; $\Delta G^{\ominus}_{1573\,K} = -13.9$ kJ mol^{-1}; $\Delta G^{\ominus}_{298\,K} = 33.5$ kJ mol^{-1}; $\Delta H^{\ominus} = 44.6$ kJ mol^{-1}; $\Delta S^{\ominus} = 37.2$ J K^{-1} mol^{-1}

Chapter 5. Solutions and Phase Equilibria

1. $\log_{10}(p/\text{N m}^{-2}) = \dfrac{-2230}{T/\text{K}} + 10.98$ (constants obtained from a plot of $\log_{10} p$ against $1/T$)
2. $\Delta H_f = 11.0$ kJ mol^{-1} 3. $\dfrac{\mathrm{d}T}{\mathrm{d}p} = -1.73 \times 10^{-6}$ K N^{-1} m^2
4. 0.705×10^5 N m^{-2}
5. $\Delta H_e = 28.5$ kJ mol^{-1}; normal boiling point = 34.6°C
6. $\Delta H_e = 30.7$ kJ mol^{-1}
7. (a) $\Delta H_{e,\,298\,K} = 60.7$ kJ mol^{-1}
    (b) $\Delta H_{e,630\,K} = 60.7$ kJ mol^{-1}; $\Delta S_{e,\,630\,K} = 96$ J K^{-1} mol^{-1}
8. 11 cm^3 (s.t.p.)
9. $x(SiCl_4)$ in liquid $= 0.29$; $x(SiCl_4)$ in initial condensate $= 0.435$
10. p (for 0.25 mole fraction n-hexane) $= 0.35 \times 10^5$ N m^{-2}; $x(n$-hexane) in vapour for lst evaporation $= 0.486$; $x(n$-hexane) in vapour for 2nd evaporation $= 0.729$
11. (a) For H_2O, $p = xp^{\bullet}$ for $0.93 < x < 1.00$; for n-C_3H_7OH, $p = xp^{\bullet}$ for $0.83 < x < 1.00$

(b) $x(H_2O) = 0.625,$ $x(n\text{-}C_3H_7OH) = 0.375$
(c) $x(H_2O) = 0.907,$ $x(n\text{-}C_3H_7OH) = 0.093$
12. 571 g   13. (a) 9.09 g   (b) 9.96 g
14. $K_c = 960 \text{ mol}^{-1} \text{ dm}^3$ 15. $K_a = 0.078$
16. $K_f = 1.86 \text{ K mol}^{-1} \text{ kg}$ 17. M_r (nitrobenzene) = 125
18. M_r (triphenylamine) = 244   19. $y = 1.64$ g
20. $p = 0.219\,56 \times 10^5 \text{ N m}^{-2},$ $M_r = 621$
21. (a) 393°C, (b) Antimony (28 mole per cent) (all of lead is still in solution) (c) 61.7 mole per cent
22. An addition compound, $(C_6H_5)_2CO \cdot (C_6H_5)_2NH$, with a congruent melting point of 40°C is formed.

Chapter 6. Reaction Kinetics

1. $5.92 \times 10^{-5} \text{ s}^{-1}$
2. First order, $k = 1.1 \times 10^{-3} \text{ s}^{-1}$; $t_{0.5} = 630$ s
3. $1.07 \times 10^{-4} \text{ s}^{-1}$ 4. 8050 s
5. $k = 3.45 \times 10^{-3} \text{ s}^{-1}$, $t_{0.5} = 200$ s
6. $1.54 \times 10^{-4} \text{ dm}^3 \text{ mol}^{-1} \text{ s}^{-1}$
7. Second order, $k = 1.67 \times 10^{-3} \text{ dm}^3 \text{ mol}^{-1} \text{ s}^{-1}$
8. $1.08 \times 10^{-2} \text{ dm}^3 \text{ mol}^{-1} \text{ s}^{-1}$
9. Second order, $k = 0.039 \text{ dm}^3 \text{ mol}^{-1} \text{ s}^{-1}$
10. Second order, $k = 1.11 \times 10^{-2} \text{ dm}^3 \text{ mol}^{-1} \text{ s}^{-1}$; $t_{0.1} = 200$ s
11. $k = 8.40 \times 10^{-5} \text{ s}^{-1}$; $t_{0.5} = 8240$ s
12. Since the reaction is first order, a plot of $\log_{10}(3p_0 - 2p)$ against t (where p_0 is the initial pressure) is linear; $k = 2.56 \times 10^{-6} \text{ s}^{-1}$
13. Plot of $\log_{10}(2p_0 - p)$ against t ($p_0 =$ initial pressure) is linear indicating that the reaction is first order; $k = 4.83 \times 10^{-4} \text{ s}^{-1}$
14. Plot of $\log_{10}(2p_0 - p)$ against t ($p_0 =$ initial pressure) is linear, indicating that the reaction is first order; $k = 2.72 \times 10^{-4} \text{ s}^{-1}$
15. Plot of $\log_{10}(p_\infty - p)$ against t is not linear, therefore the reaction is not first order. Plot of $(p_\infty - p)^{-1}$ against t is linear indicating that the reaction is second order ($k = 5 \times 10^{-4} \text{ torr}^{-1} \text{ min}^{-1}$)
16. $p_{\text{final}} = 300$ torr; $t_{0.5} = 3120$ s. First order, since half-life is independent of initial pressure, indicating weak adsorption.
17. $\Delta E^{\ddagger} = 90.0 \text{ kJ mol}^{-1}$; $A = 1.1 \times 10^{12} \text{ dm}^3 \text{ mol}^{-1} \text{ s}^{-1}$
18. $\Delta E^{\ddagger} = 211 \text{ kJ mol}^{-1}$; $A = 2.5 \times 10^{13} \text{ s}^{-1}$
19. $\Delta E^{\ddagger} = 116 \text{ kJ mol}^{-1}$; $A = 4 \times 10^9 \text{ s}^{-1}$

20. $k(0°C) = 0.667$ dm^3 mol^{-1} s^{-1}; $\Delta E^{\neq} = 59.0$ kJ mol^{-1}
21. $\Delta E^{\neq} = 115$ kJ mol^{-1}; $|A = 3.7 \times 10^{12}$ cm^3 mol^{-1} s^{-1}
22. Second order. (*a*) $k(694°C) = 0.135$ dm^3 mol^{-1} s^{-1};
   $k(757°C) = 0.842$ dm^3 mol^{-1} s^{-1} (*b*) $\Delta E^{\neq} = 241$ kJ mol^{-1}
   (*c*) $A = 1.4 \times 10^{12}$ dm^3 mol^{-1} s^{-1}
23. 1340 s   24. 36.2% of the 20% solution
25. $k(25°C) = 1.32 \times 10^{-5}$ s^{-1}; $k(50°C) = 2.08 \times 10^{-4}$ s^{-1};
   $\Delta E^{\neq} = 88.4$ kJ mol^{-1}
26. 52.9 kJ mol^{-1} 27. 282 K

Chapter 7. Electrochemistry

1. $\Lambda(CuSO_4) = 1.182 \times 10^{-2}$ Ω^{-1} m^2 mol^{-1}
2. $\Lambda^{\infty}(BaCl_2) = 2.80 \times 10^{-2}$ Ω^{-1} m^2 mol^{-1}
3. $\Lambda^{\infty}(KCl) = 1.498 \times 10^{-2}$ Ω^{-1} m^2 mol^{-1}
4. $\Lambda^{\infty}(NH_4OH) = 2.710\ 5 \times 10^{-2}$ Ω^{-1} m^2 mol^{-1}
5. $\Lambda^{\infty}(KNO_3) = 1.449\ 6 \times 10^{-2}$ Ω^{-1} m^2 mol^{-1}; Λ^{∞} (NaCl)
   $= 1.264\ 8 \times 10^{-2}$ Ω^{-1} m^2 mol^{-1}
6. $u(Na^+) = 4.46 \times 10^{-8}$ m^2 s^{-1} V^{-1}; $u(Cl^-) = 6.83 \times 10^{-8}$ m^2 s^{-1} V^{-1}
7. $u(Na^+) = 5.19 \times 10^{-8}$ m^2 s^{-1} V^{-1}; $u(SO_4^{2-}) = 8.29 \times 10^{-8}$ m^2 s^{-1} V^{-1}
8. (*a*) 0.426 Ω^{-1} m^{-1} (*b*) 0.184 Ω^{-1} m^{-1} (*c*) 0.063 Ω^{-1} m^{-1}
   (*d*) 0.100 Ω^{-1} m^{-1}
   (Λ is assumed to be concentration-independent)
9. (*a*) $t(H^+) = 0.83$, $t(Cl^-) = 0.17$ (*b*) $0.096\ 3$ mol dm^{-3}
10. $t(K^+) = 0.484$ (anode data), 0.490 (cathode data)
11. $t(Ag^+) = 0.469$, $t(NO_3^-) = 0.531$
12. $t(Na^+) = 0.39$, $t(Cl^-) = 0.61$
13. $t(K^+) = 0.492$, $u(K^+) = 6.58 \times 10^{-8}$ m^2 s^{-1} V^{-1}
14. 5.25×10^{-4} mol dm^{-3} or 0.092 g dm^{-3}
15. 6.4×10^{-5} mol dm^{-3} or $0.019\ 4$ g dm^{-3}
16. $\Lambda(\frac{1}{2}SO_4^{2-}) = 8.0 \times 10^{-3}$ Ω^{-1} m^2 mol^{-1}, $K_s(CaSO_4)$
   $= 4.03 \times 10^{-6}$
17. $\gamma_{Na^+} = \gamma_{Cl^-} = \gamma_{NO_3^-} = 0.90$; $\gamma_{La^{3+}} = 0.39$
18. 4.54×10^{-2} Ω^{-1} m^{-1} 19. $K_a' = 7.0 \times 10^{-4}$ mol dm^{-3}
20. 1.02×10^{-4} Ω^{-1} m^{-1} 21. $K_a = 1.42 \times 10^{-3}$
22. (*a*) pH 3.03 (*b*) pH 3.39 23. pH 2.64
24. (*a*) $\Lambda^{\infty}(HCOOH) = 4.043 \times 10^{-2}$ Ω^{-1} m^2 mol^{-1}; $K_a' = 1.80 \times 10^{-4}$ mol dm^{-3} (*b*) pH 2.90
25. (*a*) pH 2.4 (*b*) pH 3.1 (*c*) pH 4.3
26. (*a*) pH 4.35 (*b*) pH 4.32
27. (*a*) 190 cm^3 (*b*) phenolphthalein ($pK_{In} = 9.5$)

28. $K_a = 1.7 \times 10^{-5}$
29. $\alpha = 0.016\,1$, $K_b = 3.8 \times 10^{-10}$
30. $\alpha = 1.69 \times 10^{-4}$, pH 5·54
31. (a) (i) pH 8·10 (ii) pH 8·05 (b) (i) pH 8·60 (ii) pH 8·50
32. $K_w = 0.98 \times 10^{-14}$ 33. pH 6·50 34. pH 4·40
35. 0·16 (neglecting activity coefficients, see Example 7.15, and assuming that the coloured tautomeric form of the indicator is completely ionized)
36. $\Delta H_{\text{neutralization}} = -56.8$ kJ mol^{-1}
37. $\gamma_{\text{Cl}^-} = 0.77$ (This answer can only be regarded as an estimate, because liquid junction potential cannot be completely eliminated from this type of cell, nor can it be calculated reliably and allowed for. Single ion activity coefficients cannot be determined unequivocally.)
38. 0·068 V
39. $\gamma_{\pm}(0.1$ mol kg^{-1} HCl$) = 0.80$ [$\gamma_{\pm}(0.001$ mol kg^{-1} HCl) calculated using the Debye–Hückel limiting equation]
40. $t(\text{Cu}^{2+}) = 0.415$
41. $x(\text{H}_2) = 0.46$ (H$_2$+A) electrode is the +ve pole.
42. $\Delta G^{\ominus}_{298\,\text{K}} = -212$ kJ mol^{-1}
43. $c_{\text{Sn}^{2+}}/c_{\text{Pb}^{2+}} = 3.0$; $\Delta G^{\ominus}_{298\,\text{K}} = 2.7$ kJ mol^{-1}
44. $K_w = 1.05 \times 10^{-14}$ (p$K_w = 13.977$) (This is only an approximate value for K_w, because (a) it is impossible to completely eliminate liquid junction potential, and (b) it is assumed that $\gamma_{\text{H}^+} = \gamma_{\text{Cl}^-}$ and $\gamma_{\text{Na}^+} = \gamma_{\text{OH}^-}$)
45. $K_{\text{AgBr}} = 5.3 \times 10^{-13}$; solubility $= 7.3 \times 10^{-7}$ mol kg^{-1}

Chapter 8. True-False and Multiple Choice

TRUE-FALSE

1. T	2. F	3. F	4. T	5. F
6. T	7. T	8. T	9. F	10. T
11. T	12. T	13. F	14. F	15. F
16. T	17. F	18. T	19. F	20. T
21. F	22. T	23. T	24. T	25. F
26. F	27. F	28. T	29. T	30. T
31. T	32. T	33. T	34. F	35. T
36. F	37. T	38. F	39. F	40. T
41. F	42. F	43. T	44. T	45. T
46. F	47. F	48. T	49. F	50. T
51. F	52. T	53. F	54. F	55. F
56. T	57. F	58. T	59. T	

60.	T (cations to cathode)	61. T	62. T	63. T
64.	T	65. T	66. F	67. F	68. F
69.	F	70. T			

MULTIPLE CHOICE

1.	(*b*)	2.	(*d*)	3.	(*c*)
4.	(*a*) (*b*)	5.	(*d*)	6.	(*c*)
7.	(*a*)	8.	(*b*) (*c*) (*d*) (*f*)	9.	(*c*)
10.	(*c*)	11.	(*d*)	12.	(*b*) (*d*)
13.	(*a*)	14.	(*d*)	15.	(*b*)
16.	(*d*)	17.	(*b*)	18.	(*a*) (*d*)
19.	(*b*) (*c*)	20.	(*c*)	21.	(*a*)
22.	(*b*)	23.	(*c*)	24.	(*a*) (*b*) (*f*)
25.	(*c*)	26.	(*b*)	27.	(*c*)
28.	(*c*)	29.	(*d*)	30.	(*b*)
31.	(*a*) (*d*)	32.	(*b*)	33.	(*d*)
34.	(*c*)	35.	(*a*)	36.	(*d*)
37.	(*a*) (*b*)	38.	(*c*)		

APPENDIX I

Physical Constants

Physical constant	Symbol	Value	Logarithm
molar gas constant	$R = N_A k$	8·314 3 $\text{J K}^{-1}\,\text{mol}^{-1}$	0·919 8
Avogadro constant	N_A, L	6·022 5 × 10²³ mol^{-1}	23·779 8
Boltzmann constant	$k = R/N_A$	1·380 5 × 10⁻²³ J K^{-1}	$\overline{23}$·140 1
molar volume (ideal gas, 0°C, 1 atm)		2·241 4 × 10⁻² $\text{m}^3\,\text{mol}^{-1}$	$\overline{2}$·350 5
elementary charge	$e = F/N_A$	1·602 1 × 10⁻¹⁹ C	$\overline{19}$·204 6
Faraday constant	$F = N_A e$	9·648 7 × 10⁴ C mol^{-1}	4·984 5
permittivity of a vacuum	ε_0	8·854 2 × 10⁻¹² $\text{kg}^{-1}\,\text{m}^{-3}\,\text{s}^4\,\text{A}^2$	$\overline{12}$·947 1
permeability of a vacuum	μ_0	$4\pi \times 10^{-7}$ $\text{kg m s}^{-2}\,\text{A}^{-2}$	$\overline{6}$·099 2
velocity of light in a vacuum	c_0	2·997 9 × 10⁸ m s^{-1}	8·476 8

Physical constant	Symbol	Value	Logarithm
Planck's constant	h	$6.625\ 6 \times 10^{-34}$ J s	$\overline{34}.821\ 2$
unified atomic mass constant	$m_u = \dfrac{m_a(^{12}\text{C})}{12}$	$1.660\ 4 \times 10^{-27}$ kg	$\overline{27}.220\ 2$
mass of proton	m_p	$1.672\ 5 \times 10^{-27}$ kg	$\overline{27}.223\ 4$
mass of neutron	m_n	$1.674\ 8 \times 10^{-27}$ kg	$\overline{27}.224\ 0$
mass of electron	m_e	$9.109\ 1 \times 10^{-31}$ kg	$\overline{31}.959\ 5$
standard gravitational acceleration	g	$9.806\ 6$ m s^{-2}	$0.991\ 5$

APPENDIX II

Conversion Factors

Equality	Logarithm
$0°C = 273 \cdot 15$ K	$2 \cdot 436$ 4
1 atm $= 760$ torr	$2 \cdot 880$ 8
1 atm $= 1 \cdot 013\ 25 \times 10^5$ N m^{-2}	$5 \cdot 005$ 7
1 torr $= 133 \cdot 32$ N m^{-2}	$2 \cdot 124$ 9
$\dfrac{RT \ln 10}{F} = 0 \cdot 059\ 15$ V at 25°C	$\bar{2} \cdot 772$ 0

Debye–Hückel limiting law constant, A, for aqueous
electrolyte solutions at 25°C $= 0 \cdot 509$ mol$^{-1/2}$ kg$^{1/2}$ $\bar{1} \cdot 706$ 7

$N_A hc = 0 \cdot 119\ 62$ J mol^{-1} m	$\bar{1} \cdot 077$ 8
$\dfrac{hc}{k} = 0 \cdot 014\ 388$ K m	$\bar{2} \cdot 158$ 0
$\dfrac{h}{8\pi^2 c} = B$ (rotational constant)	
$\quad = 2 \cdot 799\ 1 \times 10^{-44}$ J m^{-1} s^2	$\overline{44} \cdot 447$ 0
$\ln 10 = 2 \cdot 302$ 6	$0 \cdot 362$ 2
$\pi = 3 \cdot 141$ 59	$0 \cdot 497$ 1

Relative Atomic Masses of the Elements

Proton number of element	Name of element	Symbol	Relative atomic mass
1	hydrogen	H	1·007 97 (1)
2	helium	He	4·002 6
3	lithium	Li	6·939
4	beryllium	Be	9·012 2
5	boron	B	10·811 (3)
6	carbon	C	12·011 15 (5)
7	nitrogen	N	14·006 7
8	oxygen	O	15·999 4 (1)
9	fluorine	F	18·998 4
10	neon	Ne	20·179 (3)
11	sodium	Na	22·989 8
12	magnesium	Mg	24·305
13	aluminium	Al	26·981 5
14	silicon	Si	28·086 (1)
15	phosphorus	P	30·973 8
16	sulphur	S	32·064 (3)
17	chlorine	Cl	35·453 (1)
18	argon	Ar	39·948
19	potassium	K	39·102
20	calcium	Ca	40·08

The figure in parentheses following some of the above values of A_r indicate the uncertainty of the final quoted figure arising from natural variations in isotopic composition and experimental uncertainty, e.g. $A_r(C) = 12·011\ 15\ (5)$ means $A_r(C)\ 12·011\ 15 \pm 0·000\ 05$ Values in parenthesis are the mass mumbers of the most stable isotopes.

Proton number of element	Name of element	Symbol	Relative atomic mass
21	scandium	Sc	44·956
22	titanium	Ti	47·90
23	vanadium	V	50·942
24	chromium	Cr	51·996
25	manganese	Mn	54·938 0
26	iron	Fe	55·847 (3)
27	cobalt	Co	58·933 2
28	nickel	Ni	58·71
29	copper	Cu	63·546 (1)
30	zinc	Zn	65·37
31	gallium	Ga	69·72
32	germanium	Ge	72·59
33	arsenic	As	74·921 6
34	selenium	Se	78·96
35	bromine	Br	79·904 (1)
36	krypton	Kr	83·80
37	rubidium	Rb	85·47
38	strontium	Sr	87·62
39	yttrium	Y	88·905
40	zirconium	Zr	91·22
41	niobium	Nb	92·906
42	molybdenum	Mo	95·94
43	technetium	Tc	(99)
44	ruthenium	Ru	101·07
45	rhodium	Rh	102·905
46	palladium	Pd	106·4
47	silver	Ag	107·868 (1)
48	cadmium	Cd	112·40
49	indium	In	114·82
50	tin	Sn	118·69
51	antimony	Sb	121·75
52	tellurium	Te	127·60
53	iodine	I	126·904 4
54	xenon	Xe	131·30
55	cesium	Cs	132·905
56	barium	Ba	137·34
57	lanthanum	La	138·91
58	cerium	Ce	140·12
59	praseodymium	Pr	140·907
60	neodymium	Nd	144·24
61	promethium	Pm	(145)
62	samarium	Sm	150·35
63	europium	Eu	151·96
64	gadolinium	Gd	157·25

Proton number of element	Name of element	Symbol	Relative atomic mass
65	terbium	Tb	158·924
66	dysprosium	Dy	162·50
67	holmium	Ho	164·930
68	erbium	Er	167·26
69	thulium	Tm	168·934
70	ytterbium	Yb	173·04
71	lutetium	Lu	174·97
72	hafnium	Hf	178·49
73	tantalum	Ta	180·948
74	tungsten	W	183·85
75	rhenium	Re	186·2
76	osmium	Os	190·2
77	iridium	Ir	192·2
78	platinum	Pt	195·09
79	gold	Au	196·967
80	mercury	Hg	200·59
81	thallium	Tl	204·37
82	lead	Pb	207·19
83	bismuth	Bi	208·980
84	polonium	Po	(210)
85	astatine	At	(210)
86	radon	Rn	(222)
87	francium	Fr	(223)
88	radium	Ra	(226)
89	actinium	Ac	(227)
90	thorium	Th	232·038
91	protactinium	Pa	(231)
92	uranium	U	238·03
93	neptunium	Np	(237)
94	plutonium	Pu	(242)
95	americium	Am	(243)
96	curium	Cm	(247)
97	berkelium	Bk	(249)
98	californium	Cf	(251)
99	einsteinium	Es	(254)

Proton number of element	Name of element	Symbol	Relative atomic mass
100	fermium	Fm	(253)
101	mendelevium	Md	(256)
102	nobelium	No	(254)
103	lawrencium	Lr	—

LOGARITHMS

	0	1	2	3	4	5	6	7	8	9	1	2	3	4	5	6	7	8	9
10	0000	0043	0086	0128	0170	0212	0253	0294	0334	0374	4	8	12	17	21	25	29	33	37
11	0414	0453	0492	0531	0569	0607	0645	0682	0719	0755	4	8	11	15	19	23	26	30	34
12	0792	0828	0864	0899	0934	0969	1004	1038	1072	1106	3	7	10	14	17	21	24	28	31
13	1139	1173	1206	1239	1271	1303	1335	1367	1399	1430	3	6	10	13	16	19	23	26	29
14	1461	1492	1523	1553	1584	1614	1644	1673	1703	1732	3	6	9	12	15	18	21	24	27
15	1761	1790	1818	1847	1875	1903	1931	1959	1987	2014	3	6	8	11	14	17	20	22	25
16	2041	2068	2095	2122	2148	2175	2201	2227	2253	2279	3	5	8	11	13	16	18	21	24
17	2304	2330	2355	2380	2405	2430	2455	2480	2504	2529	2	5	7	10	12	15	17	20	22
18	2553	2577	2601	2625	2648	2672	2695	2718	2742	2765	2	5	7	9	12	14	16	19	21
19	2788	2810	2833	2856	2878	2900	2923	2945	2967	2989	2	4	7	9	11	13	16	18	20
20	3010	3032	3054	3075	3096	3118	3139	3160	3181	3201	2	4	6	8	11	13	15	17	19
21	3222	3243	3263	3284	3304	3324	3345	3365	3385	3404	2	4	6	8	10	12	14	16	18
22	3424	3444	3464	3483	3502	3522	3541	3560	3579	3598	2	4	6	8	10	12	14	15	17
23	3617	3636	3655	3674	3692	3711	3729	3747	3766	3784	2	4	6	7	9	11	13	15	17
24	3802	3820	3838	3856	3874	3892	3909	3927	3945	3962	2	4	5	7	9	11	12	14	16
25	3979	3997	4014	4031	4048	4065	4082	4099	4116	4133	2	3	5	7	9	10	12	14	15
26	4150	4166	4183	4200	4216	4232	4249	4265	4281	4298	2	3	5	7	8	10	11	13	15
27	4314	4330	4346	4362	4378	4393	4409	4425	4440	4456	2	3	5	6	8	9	11	13	14
28	4472	4487	4502	4518	4533	4548	4564	4579	4594	4609	2	3	5	6	8	9	11	12	14
29	4624	4639	4654	4669	4683	4698	4713	4728	4742	4757	1	3	4	6	7	9	10	12	13
30	4771	4786	4800	4814	4829	4843	4857	4871	4886	4900	1	3	4	6	7	9	10	11	13
31	4914	4928	4942	4955	4969	4983	4997	5011	5024	5038	1	3	4	6	7	8	10	11	12
32	5051	5065	5079	5092	5105	5119	5132	5145	5159	5172	1	3	4	5	7	8	9	11	12
33	5185	5198	5211	5224	5237	5250	5263	5276	5289	5302	1	3	4	5	6	8	9	10	12
34	5315	5328	5340	5353	5366	5378	5391	5403	5416	5428	1	3	4	5	6	8	9	10	11
35	5441	5453	5465	5478	5490	5502	5514	5527	5539	5551	1	2	4	5	6	7	9	10	11
36	5563	5575	5587	5599	5611	5623	5635	5647	5658	5670	1	2	4	5	6	7	8	10	11
37	5682	5694	5705	5717	5729	5740	5752	5763	5775	5786	1	2	3	5	6	7	8	9	10
38	5798	5809	5821	5832	5843	5855	5866	5877	5888	5899	1	2	3	5	6	7	8	9	10
39	5911	5922	5933	5944	5955	5966	5977	5988	5999	6010	1	2	3	4	5	7	8	9	10
40	6021	6031	6042	6053	6064	6075	6085	6096	6107	6117	1	2	3	4	5	6	8	9	10
41	6128	6138	6149	6160	6170	6180	6191	6201	6212	6222	1	2	3	4	5	6	7	8	9
42	6232	6243	6253	6263	6274	6284	6294	6304	6314	6325	1	2	3	4	5	6	7	8	9
43	6335	6345	6355	6365	6375	6385	6395	6405	6415	6425	1	2	3	4	5	6	7	8	9
44	6435	6444	6454	6464	6474	6484	6493	6503	6513	6522	1	2	3	4	5	6	7	8	9
45	6532	6542	6551	6561	6571	6580	6590	6599	6609	6618	1	2	3	4	5	6	7	8	9
46	6628	6637	6646	6656	6665	6675	6684	6693	6702	6712	1	2	3	4	5	6	7	7	8
47	6721	6730	6739	6749	6758	6767	6776	6785	6794	6803	1	2	3	4	5	5	6	7	8
48	6812	6821	6830	6839	6848	6857	6866	6875	6884	6893	1	2	3	4	4	5	6	7	8
49	6902	6911	6920	6928	6937	6946	6955	6964	6972	6981	1	2	3	4	4	5	6	7	8
50	6990	6998	7007	7016	7024	7033	7042	7050	7059	7067	1	2	3	3	4	5	6	7	8
51	7076	7084	7093	7101	7110	7118	7126	7135	7143	7152	1	2	3	3	4	5	6	7	8
52	7160	7168	7177	7185	7193	7202	7210	7218	7226	7235	1	2	2	3	4	5	6	7	7
53	7243	7251	7259	7267	7275	7284	7292	7300	7308	7316	1	2	2	3	4	5	6	6	7
54	7324	7332	7340	7348	7356	7364	7372	7380	7388	7396	1	2	2	3	4	5	6	6	7

LOGARITHMS

	0	1	2	3	4	5	6	7	8	9	1	2	3	4	5	6	7	8	9
55	7404	7412	7419	7427	7435	7443	7451	7459	7466	7474	1	2	2	3	4	5	5	6	7
56	7482	7490	7497	7505	7513	7520	7528	7536	7543	7551	1	2	2	3	4	5	5	6	7
57	7559	7566	7574	7582	7589	7597	7604	7612	7619	7627	1	2	2	3	4	5	5	6	7
58	7634	7642	7649	7657	7664	7672	7679	7686	7694	7701	1	1	2	3	4	4	5	6	7
59	7709	7716	7723	7731	7738	7745	7752	7760	7767	7774	1	1	2	3	4	4	5	6	7
60	7782	7789	7796	7803	7810	7818	7825	7832	7839	7846	1	1	2	3	4	4	5	6	6
61	7853	7860	7868	7875	7882	7889	7896	7903	7910	7917	1	1	2	3	4	4	5	6	6
62	7924	7931	7938	7945	7952	7959	7966	7973	7980	7987	1	1	2	3	3	4	5	6	6
63	7993	8000	8007	8014	8021	8028	8035	8041	8048	8055	1	1	2	3	3	4	5	5	6
64	8062	8069	8075	8082	8089	8096	8102	8109	8116	8122	1	1	2	3	3	4	5	5	6
65	8129	8136	8142	8149	8156	8162	8169	8176	8182	8189	1	1	2	3	3	4	5	5	6
66	8195	8202	8209	8215	8222	8228	8235	8241	8248	8254	1	1	2	3	3	4	5	5	6
67	8261	8267	8274	8280	8287	8293	8299	8306	8312	8319	1	1	2	3	3	4	5	5	6
68	8325	8331	8338	8344	8351	8357	8363	8370	8376	8382	1	1	2	3	3	4	4	5	6
69	8388	8395	8401	8407	8414	8420	8426	8432	8439	8445	1	1	2	2	3	4	4	5	6
70	8451	8457	8463	8470	8476	8482	8488	8494	8500	8506	1	1	2	2	3	4	4	5	6
71	8513	8519	8525	8531	8537	8543	8549	8555	8561	8567	1	1	2	2	3	4	4	5	5
72	8573	8579	8585	8591	8597	8603	8609	8615	8621	8627	1	1	2	2	3	4	4	5	5
73	8633	8639	8645	8651	8657	8663	8669	8675	8681	8686	1	1	2	2	3	4	4	5	5
74	8692	8698	8704	8710	8716	8722	8727	8733	8739	8745	1	1	2	2	3	4	4	5	5
75	8751	8756	8762	8768	8774	8779	8785	8791	8797	8802	1	1	2	2	3	3	4	5	5
76	8808	8814	8820	8825	8831	8837	8842	8848	8854	8859	1	1	2	2	3	3	4	5	5
77	8865	8871	8876	8882	8887	8893	8899	8904	8910	8915	1	1	2	2	3	3	4	4	5
78	8921	8927	8932	8938	8943	8949	8954	8960	8965	8971	1	1	2	2	3	3	4	4	5
79	8976	8982	8987	8993	8998	9004	9009	9015	9020	9025	1	1	2	2	3	3	4	4	5
80	9031	9036	9042	9047	9053	9058	9063	9069	9074	9079	1	1	2	2	3	3	4	4	5
81	9085	9090	9096	9101	9106	9112	9117	9122	9128	9133	1	1	2	2	3	3	4	4	5
82	9138	9143	9149	9154	9159	9165	9170	9175	9180	9186	1	1	2	2	3	3	4	4	5
83	9191	9196	9201	9206	9212	9217	9222	9227	9232	9238	1	1	2	2	3	3	4	4	5
84	9243	9248	9253	9258	9263	9269	9274	9279	9284	9289	1	1	2	2	3	3	4	4	5
85	9294	9299	9304	9309	9315	9320	9325	9330	9335	9340	1	1	2	2	3	3	4	4	5
86	9345	9350	9355	9360	9365	9370	9375	9380	9385	9390	1	1	2	2	3	3	4	4	5
87	9395	9400	9405	9410	9415	9420	9425	9430	9435	9440	0	1	1	2	2	3	3	4	4
88	9445	9450	9455	9460	9465	9469	9474	9479	9484	9489	0	1	1	2	2	3	3	4	4
89	9494	9499	9504	9509	9513	9518	9523	9528	9533	9538	0	1	1	2	2	3	3	4	4
90	9542	9547	9552	9557	9562	9566	9571	9576	9581	9586	0	1	1	2	2	3	3	4	4
91	9590	9595	9600	9605	9609	9614	9619	9624	9628	9633	0	1	1	2	2	3	3	4	4
92	9638	9643	9647	9652	9657	9661	9666	9671	9675	9680	0	1	1	2	2	3	3	4	4
93	9685	9689	9694	9699	9703	9708	9713	9717	9722	9727	0	1	1	2	2	3	3	4	4
94	9731	9736	9741	9745	9750	9754	9759	9763	9768	9773	0	1	1	2	2	3	3	4	4
95	9777	9782	9786	9791	9795	9800	9805	9809	9814	9818	0	1	1	2	2	3	3	4	4
96	9823	9827	9832	9836	9841	9845	9850	9854	9859	9863	0	1	1	2	2	3	3	4	4
97	9868	9872	9877	9881	9886	9890	9894	9899	9903	9908	0	1	1	2	2	3	3	4	4
98	9912	9917	9921	9926	9930	9934	9939	9943	9948	9952	0	1	1	2	2	3	3	4	4
99	9956	9961	9965	9969	9974	9978	9983	9987	9991	9996	0	1	1	2	2	3	3	3	4

154 *Antilogarithms*

ANTILOGARITHMS

	0	1	2	3	4	5	6	7	8	9	1	2	3	4	5	6	7	8	9
·00	1000	1002	1005	1007	1009	1012	1014	1016	1019	1021	0	0	1	1	1	1	2	2	2
·01	1023	1026	1028	1030	1033	1035	1038	1040	1042	1045	0	0	1	1	1	1	2	2	2
·02	1047	1050	1052	1054	1057	1059	1062	1064	1067	1069	0	0	1	1	1	1	2	2	2
·03	1072	1074	1076	1079	1081	1084	1086	1089	1091	1094	0	0	1	1	1	1	2	2	2
·04	1095	1099	1102	1104	1107	1109	1112	1114	1117	1119	0	1	1	1	1	2	2	2	2
·05	1122	1125	1127	1130	1132	1135	1138	1140	1143	1146	0	1	1	1	1	2	2	2	2
·06	1148	1151	1153	1156	1159	1161	1164	1167	1169	1172	0	1	1	1	1	2	2	2	2
·07	1175	1178	1180	1183	1186	1189	1191	1194	1197	1199	0	1	1	1	1	2	2	2	2
·08	1202	1205	1208	1211	1213	1216	1219	1222	1225	1227	0	1	1	1	1	2	2	2	3
·09	1230	1233	1236	1239	1242	1245	1247	1250	1253	1256	0	1	1	1	1	2	2	2	3
·10	1259	1262	1265	1268	1271	1274	1276	1279	1282	1285	0	1	1	1	1	2	2	2	3
·11	1288	1291	1294	1297	1300	1303	1306	1309	1312	1315	0	1	1	1	2	2	2	2	3
·12	1318	1321	1324	1327	1330	1334	1337	1340	1343	1346	0	1	1	1	2	2	2	2	3
·13	1349	1352	1355	1358	1361	1365	1368	1371	1374	1377	0	1	1	1	2	2	2	3	3
·14	1380	1384	1387	1390	1393	1396	1400	1403	1406	1409	0	1	1	1	2	2	2	3	3
·15	1413	1416	1419	1422	1426	1429	1432	1435	1439	1442	0	1	1	1	2	2	2	3	3
·16	1445	1449	1452	1455	1459	1462	1466	1469	1472	1476	0	1	1	1	2	2	2	3	3
·17	1479	1483	1486	1489	1493	1496	1500	1503	1507	1510	0	1	1	1	2	2	2	3	3
·18	1514	1517	1521	1524	1528	1531	1535	1538	1542	1545	0	1	1	1	2	2	2	3	3
·19	1549	1552	1556	1560	1563	1567	1570	1574	1578	1581	0	1	1	1	2	2	3	3	3
·20	1585	1589	1592	1596	1600	1603	1607	1611	1614	1618	0	1	1	1	2	2	3	3	3
·21	1622	1626	1629	1633	1637	1641	1644	1648	1652	1656	0	1	1	2	2	2	3	3	3
·22	1660	1663	1667	1671	1675	1679	1683	1687	1690	1694	0	1	1	2	2	2	3	3	3
·23	1698	1702	1706	1710	1714	1718	1722	1726	1730	1734	0	1	1	2	2	2	3	3	4
·24	1738	1742	1746	1750	1754	1758	1762	1766	1770	1774	0	1	1	2	2	2	3	3	4
·25	1778	1782	1786	1791	1795	1799	1803	1807	1811	1816	0	1	1	2	2	2	3	3	4
·26	1820	1824	1828	1832	1837	1841	1845	1848	1854	1858	0	1	1	2	2	3	3	3	4
·27	1862	1866	1871	1875	1879	1884	1888	1892	1897	1901	0	1	1	2	2	3	3	3	4
·28	1905	1910	1914	1919	1923	1928	1932	1936	1941	1945	0	1	1	2	2	3	3	4	4
·29	1950	1954	1959	1963	1968	1972	1977	1982	1986	1991	0	1	1	2	2	3	3	4	4
·30	1995	2000	2004	2009	2014	2018	2023	2028	2032	2037	0	1	1	2	2	3	3	4	4
·31	2042	2046	2051	2056	2061	2065	2070	2075	2080	2084	0	1	1	2	2	3	3	4	4
·32	2089	2094	2099	2104	2109	2113	2118	2123	2128	2133	0	1	1	2	2	3	3	4	4
·33	2138	2143	2148	2153	2158	2163	2168	2173	2178	2183	0	1	1	2	2	3	3	4	4
·34	2188	2193	2198	2203	2208	2213	2218	2223	2228	2234	1	1	2	2	3	3	4	4	5
·35	2239	2244	2249	2254	2259	2265	2270	2275	2280	2286	1	1	2	2	3	3	4	4	5
·36	2291	2296	2301	2307	2312	2317	2323	2328	2333	2339	1	1	2	2	3	3	4	4	5
·37	2344	2350	2355	2360	2366	2371	2377	2382	2388	2393	1	1	2	2	3	3	4	4	5
·38	2399	2404	2410	2415	2421	2427	2432	2438	2443	2449	1	1	2	2	3	3	4	4	5
·39	2455	2460	2466	2472	2477	2483	2489	2495	2500	2506	1	1	2	2	3	3	4	5	5
·40	2512	2518	2523	2529	2535	2541	2547	2553	2559	2564	1	1	2	2	3	4	4	5	5
·41	2570	2576	2582	2588	2594	2600	2606	2612	2618	2624	1	1	2	2	3	4	4	5	5
·42	2630	2636	2642	2649	2655	2661	2667	2673	2679	2685	1	1	2	2	3	4	4	5	6
·43	2692	2698	2704	2710	2716	2723	2729	2735	2742	2748	1	1	2	3	3	4	4	5	6
·44	2754	2761	2767	2773	2780	2786	2793	2799	2805	2812	1	1	2	3	3	4	4	5	6
·45	2818	2825	2831	2838	2844	2851	2858	2864	2871	2877	1	1	2	3	3	4	5	5	6
·46	2884	2891	2897	2904	2911	2917	2924	2931	2938	2944	1	1	2	3	3	4	5	5	6
·47	2951	2958	2965	2972	2979	2985	2992	2999	3006	3013	1	1	2	3	3	4	5	5	6
·48	3020	3027	3034	3041	3048	3055	3062	3069	3076	3083	1	1	2	3	4	4	5	6	6
·49	3090	3097	3105	3112	3119	3126	3133	3141	3148	3155	1	1	2	3	4	4	5	6	6

ANTILOGARITHMS

	0	1	2	3	4	5	6	7	8	9	1	2	3	4	5	6	7	8	9
·50	3162	3170	3177	3184	3192	3199	3206	3214	3221	3228	1	1	2	3	4	4	5	6	7
·51	3236	3243	3251	3258	3266	3273	3281	3289	3296	3304	1	2	2	3	4	5	5	6	7
·52	3311	3319	3327	3334	3342	3350	3357	3365	3373	3381	1	2	2	3	4	5	5	6	7
·53	3388	3396	3404	3412	3420	3428	3436	3443	3451	3459	1	2	2	3	4	5	6	6	7
·54	3467	3475	3483	3491	3499	3508	3516	3524	3532	3540	1	2	2	3	4	5	6	6	7
·55	3548	3556	3565	3573	3581	3589	3597	3606	3614	3622	1	2	2	3	4	5	6	7	7
·56	3631	3639	3648	3656	3664	3673	3681	3690	3698	3707	1	2	3	3	4	5	6	7	8
·57	3715	3724	3733	3741	3750	3758	3767	3776	3784	3793	1	2	3	3	4	5	6	7	8
·58	3802	3811	3819	3828	3837	3846	3855	3864	3873	3882	1	2	3	4	4	5	6	7	8
·59	3890	3899	3908	3917	3926	3936	3945	3954	3963	3972	1	2	3	4	5	5	6	7	8
·60	3981	3990	3999	4009	4018	4027	4036	4046	4055	4064	1	2	3	4	5	6	6	7	8
·61	4074	4083	4093	4102	4111	4121	4130	4140	4150	4159	1	2	3	4	5	6	7	8	9
·62	4169	4178	4188	4198	4207	4217	4227	4236	4246	4256	1	2	3	4	5	6	7	8	9
·63	4266	4276	4285	4295	4305	4315	4325	4335	4345	4355	1	2	3	4	5	6	7	8	9
·64	4365	4375	4385	4395	4406	4416	4426	4436	4446	4457	1	2	3	4	5	6	7	8	9
·65	4467	4477	4487	4498	4508	4519	4529	4539	4550	4560	1	2	3	4	5	6	7	8	9
·66	4571	4581	4592	4603	4613	4624	4634	4645	4656	4667	1	2	3	4	5	6	7	9	10
·67	4677	4688	4699	4710	4721	4732	4742	4753	4764	4775	1	2	3	4	5	7	8	9	10
·68	4786	4797	4808	4819	4831	4842	4853	4864	4875	4887	1	2	3	4	6	7	8	9	10
·69	4898	4909	4920	4932	4943	4955	4966	4977	4989	5000	1	2	3	5	6	7	8	9	10
·70	5012	5023	5035	5047	5058	5070	5082	5093	5105	5117	1	2	4	5	6	7	8	9	11
·71	5129	5140	5152	5164	5176	5188	5200	5212	5224	5236	1	2	4	5	6	7	8	10	11
·72	5248	5260	5272	5284	5297	5309	5321	5333	5346	5358	1	2	4	5	6	7	9	10	11
·73	5370	5383	5395	5408	5420	5433	5445	5458	5470	5483	1	3	4	5	6	8	9	10	11
·74	5495	5508	5521	5534	5546	5559	5572	5585	5598	5610	1	3	4	5	6	8	9	10	12
·75	5623	5636	5649	5662	5675	5689	5702	5715	5728	5741	1	3	4	5	7	8	9	10	12
·76	5754	5768	5781	5794	5808	5821	5834	5848	5861	5875	1	3	4	5	7	8	9	11	12
·77	5888	5902	5916	5929	5943	5957	5970	5984	5998	6012	1	3	4	5	7	8	10	11	12
·78	6026	6039	6053	6067	6081	6095	6109	6124	6138	6152	1	3	4	6	7	8	10	11	13
·79	6166	6180	6194	6209	6223	6237	6252	6266	6281	6295	1	3	4	6	7	9	10	11	13
·80	6310	6324	6339	6353	6368	6383	6397	6412	6427	6442	1	3	4	6	7	9	10	12	13
·81	6457	6471	6486	6501	6516	6531	6546	6561	6577	6592	2	3	5	6	8	9	11	12	14
·82	6607	6622	6637	6653	6668	6683	6699	6714	6730	6745	2	3	5	6	8	9	11	12	14
·83	6761	6776	6792	6808	6823	6839	6855	6871	6887	6902	2	3	5	6	8	9	11	13	14
·84	6918	6934	6950	6966	6982	6998	7015	7031	7047	7063	2	3	5	6	8	10	11	13	15
·85	7079	7096	7112	7129	7145	7161	7178	7194	7211	7228	2	3	5	7	8	10	12	13	15
·86	7244	7261	7278	7295	7311	7328	7345	7362	7379	7396	2	3	5	7	8	10	12	13	15
·87	7413	7430	7447	7464	7482	7499	7516	7534	7551	7568	2	4	5	7	9	10	12	14	16
·88	7586	7603	7621	7638	7656	7674	7691	7709	7727	7745	2	4	5	7	9	11	12	14	16
·89	7762	7780	7798	7816	7834	7852	7870	7889	7907	7925	2	4	5	7	9	11	13	14	16
·90	7943	7962	7980	7998	8017	8035	8054	8072	8091	8110	2	4	6	7	9	11	13	15	17
·91	8128	8147	8166	8185	8204	8222	8241	8260	8279	8299	2	4	6	8	9	11	13	15	17
·92	8318	8337	8356	8375	8395	8414	8433	8453	8472	8492	2	4	6	8	10	12	14	15	17
·93	8511	8531	8551	8570	8590	8610	8630	8650	8670	8690	2	4	6	8	10	12	14	16	18
·94	8710	8730	8750	8770	8790	8810	8831	8851	8872	8892	2	4	6	8	10	12	14	16	18
·95	8913	8933	8954	8974	8995	9016	9034	9057	9078	9099	2	4	6	8	10	12	15	17	19
·96	9120	9141	9162	9183	9204	9226	9247	9268	9290	9311	2	4	6	8	11	13	15	17	19
·97	9333	9354	9376	9397	9419	9441	9462	9484	9506	9528	2	4	7	9	11	13	15	17	20
·98	9550	9572	9594	9616	9638	9661	9683	9705	9727	9750	2	4	7	9	11	13	16	18	20
·99	9772	9795	9817	9840	9863	9886	9908	9931	9954	9977	2	5	7	9	11	14	16	18	20

Index